Lecture Notes in Mathematics

Edited by A. Dold and B. Eckmann

W9-ABN-155

621

Robert M. Kauffman
Thomas T. Read
Anton Zettl

The Deficiency Index Problem for Powers of Ordinary Differential Expressions

Springer-Verlag
Berlin Heidelberg New York 1977

Authors

Robert M. Kauffman
Thomas T. Read
Department of Mathematics
and Computer Science
Western Washington University
Bellingham, WA 98225/USA

Anton Zettl
Department of Mathematical Sciences
Northern Illinois University
DeKalb, IL 60115/USA

AMS Subject Classifications (1970): 34 B 20, 47 E 05

ISBN 3-540-08523-8 Springer-Verlag Berlin Heidelberg New York
ISBN 0-387-08523-8 Springer-Verlag New York Heidelberg Berlin

Printing and binding: Beltz Offsetdruck, Hemsbach/Bergstr.
2140/3140-543210

Math

S-tpr

Preface

These Notes were started when all three authors were visiting the University of Dundee in the Spring of 1975. We are grateful to the University of Dundee and in particular to Professor W.N. Everitt for the opportunity of this visit and for his suggestion that we undertake this project. In addition the last named author would like to thank the British Science Research Council for their financial support. Also we would like to thank Ms. Barbara Bruns and Ms. Sara Clayton for their careful typing of a difficult manuscript.

Anton Zettl

Department of Mathematical Sciences

Northern Illinois University

DeKalb, Illinois 60115 U.S.A.

R. M. Kauffman and T. T. Read

Department of Mathematics and Computer Science

Western Washington University

Bellingham, Washington, U.S.A.

Contents

Introduction

The deficiency index d of an ordinary linear symmetric (formally self-adjoint) differential expression with real coefficients

$$(0.1) \qquad My = \sum_{j=0}^{n} (p_j y^{(j)})^{(j)} \quad \text{on} \quad [0,\infty)$$

can be interpreted as the number of linearly independent solutions of the equation

$$(0.2) \qquad My = \lambda y, \quad \text{Im } \lambda \neq 0$$

which lie in $L^2(0,\infty)$. This integer d is independent of the complex number λ (provided Im $\lambda \neq 0$) and so depends only on the expression M which in turn is completely determined by the coefficients p_j: $d = d(M) = d(p_0, p_1, \ldots, p_n)$.

The deficiency index problem is the problem of determining d in terms of the p_j. This problem dates back at least to a celebrated paper of H. Weyl in 1910 and is currently receiving a lot of attention in the literature: a large number and wide variety of new results have been discovered in the last ten years.

If the coefficients p_j are sufficiently differentiable, powers of M can be formed in the natural way: $M^2 y = M(My)$, $M^k y = M(M^{k-1} y)$. These powers M^k are again symmetric expressions of type (0.1). Hence the coefficients p_j determine not only the deficiency index of M but a whole sequence of deficiency indices $d(M)$, $d(M^2)$, $d(M^3)$,...

In these Notes we discuss the determination of these sequences. A complete characterization of the possible such sequences, for the case with real coefficients p_j, is obtained. Also it is shown that some of the well known sufficient conditions for the limit point case, i.e. the case when the deficiency index is minimal, also imply that all powers are in the limit point case. Thus the well known Levinson condition for second order expressions to be in the limit point case actually implies that all powers are limit point. In fact, as we shall see below, the Levinson condition can be strengthened substantially and this stronger version implies that all powers are limit point. This strong form of the Levinson condition includes many of the recently obtained second order limit-point criteria. Among these are the interval criteria of Hartman, Eastham, Atkinson-Evans, Ismagilo as well as the integral conditions of Atkinson and Brinck.

For the higher order cases the well known condition of Hinton is strengthened so as to yield strong interval type criteria. This strong version of Hinton's condition implies, in addition, that all powers are limit-point.

Below Theorem a.b.c refers to Theorem c of section b in chapter a, Corollary b.c refers to Corollary c of section b in the same chapter.

CHAPTER 1

FUNCTIONAL ANALYTIC PRELIMINARIES

Here we merely list, for the convenience of the reader, some results from Functional Analysis to be used later. For proofs the reader is referred to the books by Dunford and Schwartz V. I and II [21] and Goldberg [63].

1. **General Operator Theory.**

1.1 Theorem. (Open mapping theorem) Let A be a bounded operator from one Banach space into another. Suppose A is 1-1 and onto. Then A^{-1} is bounded.

1.2 Definition. Let X and Y be Banach spaces. An operator A from X to Y is said to be closed if its graph is a closed subset of $X \times Y$.

1.3 Theorem. (Closed graph theorem) Let A be a closed operator whose domain is a Banach space and whose range is contained in a Banach space. Then A is bounded.

1.4 Theorem. Let A be a bounded operator from a B-space into a B-space and let A* be its adjoint. Then $||A*|| = ||A||$.

1.5 Theorem. Let A be a closed, densely defined operator from a Hilbert space H_1 into a Hilbert space H_2. Suppose the range of A, R(A) is closed. Then R(A) is the orthogonal complement of the null space of A*.

1.6 Theorem. Suppose A is a closed, densely defined operator from a Hilbert space H_1 into a Hilbert space H_2. Then R(A) is closed if and only if R(A*) is closed.

1.7 Definition. The index of an operator A is defined as the nullity of A minus the deficiency of R(A) whenever at least one of these is finite.

1.8 Theorem. If B is an extension of A and the dimension of the quotient space domain B/domain A is n, then index B = index A + n.

1.9 Theorem. Let S be a closed subspace of a Hilbert space and let F be a finite dimensional subspace. Then S + F is a closed subspace.

1.10 Corollary. Let A and B be operators from a Hilbert space into a Hilbert space. Suppose B is a finite dimensional extension of A i.e. B agrees with A on D(A) and the quotient space domain B/domain A is finite dimensional. If

A is closed then B is closed.

1.11 Definition. An operator is closeable if it has a closed extension.

1.12 Theorem. Let A be an operator from a Hilbert space H_1 into a Hilbert space H_2. Suppose D(A*) is dense in H_2. Then A is closeable.

2. Symmetric Operators

2.1 Definition. A densely defined operator A on a Hilbert space is said to be symmetric if (Ax,y) = (x,Ay) for all x, y in the domain of A.

Note that a symmetric operator A is always closeable since $A \subseteq A*$ and the adjoint operator is closed.

2.2 Definition. Let A and B be operators. Then the sum A + B is defined as follows:

(a) D(A+B) = D(A) ∩ D(B) and

(b) (A+B)x = Ax + Bx for all x in D(A+B)

Similarly the product operator AB is defined as:

(a) D(AB) = {x ε D(B) | Bx ε D(A)} and

(b) (AB)x = A(Bx) for all x in D(AB)

For a scalar c and an operator A we define cA as D(cA) = D(A) and (cA)x = cAx for all x in D(A).

2.3 Theorem. If A is symmetric, then A + cI has no non trivial null space unless c is real.

2.4 Definition. Let A be a symmetric operator on a Hilbert space. Then the positive deficiency index of A, denote by $d_+(A)$ is the nullity of A* + iI and the negative deficiency index of A, $d_-(A)$, is the nullity of A* - iI where $i = \sqrt{-1}$.

2.5 Theorem. Let A be a symmetric operator on a Hilbert space. Then, for any complex number λ with positive imaginary part, the nullity of A* + λI = $d_+(A)$ and nullity of A* - λI = $d_-(A)$. Furthermore if A has closed range $d_+(A) = d_-(A)$ = nullity of A*.

2.6 Theorem. (von Neumann's formulas). Let A be a symmetric operator on a Hilbert space. Then the domain of A* is the direct sum of: the domain of A, the nullspace of $A^* - iI$ and the null space of $A^* + iI$.

<u>2.7 Theorem</u>. Suppose A is a symmetric operator on a Hilbert space H. Then A has self-adjoint extensions in H if and only if $d_+(A) = d_-(A)$.

<u>2.8 Theorem</u>. Let A be a symmetric operator on a Hilbert space H. Then A has self-adjoint extensions in $H \times H$.

<u>2.9 Theorem</u>. If A is a closed symmetric operator and B is a bounded symmetric operator, then A and $A + B$ have the same deficiency indices.

CHAPTER II

LINEAR DIFFERENTIAL OPERATORS AND THE GENERAL CLASSIFICATION
THEORY OF DEFICIENCY INDICES.

In this chapter we (1) review the general classification theory for deficiency indices of ordinary symmetric (formally self-adjoint) differential expressions and (2) develop, from basic principles, the theory of ordinary (not necessarily symmetric) differential expressions in the setting of $L^2(0,\infty)$. Thus the reader who is familiar with the elementary theory of differential operators may wish to go on to chapter III possibly after perusing through the review of the classification theory for the deficiency index.

It is possible to take the point of view that the theory of deficiency indices is really about the square integrable solution space of any - not necessarily symmetric-linear differential equation. This is the approach we take here. In particular the limit-point or "limit-n" theory is extended to the non symmetric case. In the process many of the results stated in the review are proven. Several of the proofs given are simpler than the usual ones. The approach taken is tailored for use in later chapters.

It should be remarked that some of our results pertaining to the extension of the limit point theory to the non symmetric case are used later to prove theorems about the symmetric case.

1. Symmetric differential expressions.

Let M be defined by

(1.1) $\qquad My = \sum_{j=0}^{m} p_j y^{(j)}$

where the coefficients p_j are complex valued functions defined on the interval $[0,\infty)$. The formal adjoint expression M^+ is defined by

(1.2) $\qquad M^+ y = \sum_{j=0}^{m} (-1)^j (\bar{p}_j y)^{(j)}.$

The relationship between M and M^+ is given by the Lagrange identity:

<u>1.1 Lemma</u>. Suppose $u^{(m-1)}$ and $v^{(m-1)}$ are absolutely continuous, then

(1.3) $uMv - vM^{+}u = [v, u]'$ where

(1.4) $[v, u] = \sum_{j=1}^{m} \sum_{k=1}^{j} (-1)^{k+1} v^{(j-k)} \overline{(p_j u)}^{(k-1)}$

This identity is established by an integration by parts. Integrating both sides of (1.3) yields Green's formula:

(1.5) $\int_{a}^{b} \overline{u}Mv - \int_{a}^{b} \overline{vM^{+}u} = [v, u](b) - [v, u](a).$

If $p_j \in C^j$ then the expression (1.2) for M^{+} can be written in the form (1.1). Then M^{+} and M can be compared. Thus we are led to

1.2 Definition. The expression M is symmetric or formally self-adjoint if $M = M^{+}$ i.e. if $My = M^{+}y$ for all y in C^m.

1.3 Definition. The expression M is said to be regular on $[0, \infty)$ if the leading coefficient p_m has no zero on $[0, \infty)$. Note that M is regular if and only if M^{+} is regular.

A closed form for the symmetric differential expressions (1.1) with $p_j \in C^j$ can be given:

(1.6) $My = \sum_{j=0}^{[m/2]} (-1)^j (a_j y^{(j)})^{(j)}$

$\qquad\qquad + i\sum_{j=0}^{[(m-1)/2]} \{(b_j y^{(j+1)})^{(j)} + (b_j y^{(j)})^{(j+1)}\}$

where a_j and b_j are real valued functions and $[x]$ denotes the greatest integer $\leq x$. For a proof of this representation see [21, v. 2, pp. 1289-91].

If M is real i.e. all coefficients p_j are real then the imaginary part of (1.6) vanishes. Thus all real symmetric expressions, with sufficiently smooth coefficients, must be of even order, say $m = 2n$ and have the form

(1.7) $My = \sum_{j=0}^{n} (-1)^j (p_{n-j} y^{(j)})^{(j)}$

with the coefficients p_j being real valued. For $m = 2$ i.e. $n = 1$ we get the Sturm-Liouville operator

(1.8) $My = -(py')' + qy.$

2. Deficiency Indices. In this section we give a definition of deficiency indices for ordinary symmetric differential expressions and list the basic classification results. For proofs the reader is referred to Glassman [61] for the real and to Everitt [38] for the complex case. Although Everitt considers expressions in a form different from (1.6) his analysis applies to (1.6).

For a complex number λ, let $d(\lambda)$ denote the number of linearly independent solutions of $My = \lambda y$ which lie in $L^2(0, \infty)$.

2.1 Theorem. Suppose M is given by (1.6) with

(B) $a_j, b_j \in L_{loc}$ and the leading coefficient (a_n if $m = 2n$ and b_n if $m = 2n + 1$) either positive or negative on $[0, \infty)$ and its reciprocal in L_{loc}.

Then $d(\lambda_1) = d(\lambda_2)$ if $\text{Im } \lambda_i > 0$ for $i = 1, 2$ or $\text{Im } \lambda_i < 0$ for $i = 1, 2$.

Although this theorem is proven in Naimark [97] only for real M the same technique of proof works in our setting.

The fact that $d(\lambda)$ is independent of λ for λ in the upper half plane or in the lower half plane prompts the next definition.

2.2 Definition. Let $d_+ = d(i)$ and $d_- = d(-i)$. The integers d_+, d_- are called the deficiency indices of M. These depend only on the expression M which in turn is completely determined by the coefficients a_j, b_j. If we wish to indicate the dependence of d_+ (or d_-) on M or its coefficients we write:

$d_+ = d_+(M) = d_+(a_0, a_1, \ldots, a_{[m/2]}, b_0, b_1, \ldots, b_{[(m-1)/2]})$. The deficiency index problem is the problem of determining d_+, d_- in terms of the coefficients a_j, b_j.

2.3 Theorem. If M is real i.e. given by (2.7) (with $a_n > 0$, $1/a_n$, a_0, $\ldots$, a_{n-1} in L_{loc}), then $d_+ = d_-$.

In this case we speak of the deficiency index of M and use the notation $d = d_+ = d_- = d(M)$.

Clearly the order m of the expression M is an upper bound for d_+ and d_-. What the best lower bound is and what values between these bounds actually occur is not so clear.

2.4 Theorem. Let M be given by (2.6) with coefficients satisfying the basic conditions (B).

(a) Suppose $m = 2n$. Then

$$n \le d_+, \; d_- \le 2n = m$$

(b) Suppose $m = 2n + 1 > 1$. There are two subcases here depending on whether the leading coefficient is positive or negative.

   (i) $b_{[(m-1)/2]} > 0$. Then $n \le d_+ \le 2n + 1 = m$ and $n + 1 \le d_- \le 2n + 1 = m$.

   (ii) $b_{[(m-1)/2]} < 0$. Then $n + 1 \le d_+ \le 2n + 1 = m$ and $n \le d_- \le 2n + 1 = m$.

(c) The upper index $d_+ = m$ if and only if the lower index $d_- = m$.

   In [86] Kogan and Rofe-Beketov have shown that all possible values of d_+ and d_- subject to (a), (b) and (c) above <u>and</u>

(d) $$|d_+ - d_-| \le 1$$

actually occur. Recently, Gilbert has shown that expressions M of type (1.6) exist with the property that $|d_+ - d_-| \ge 2$. (See remark on p. 21)

<u>2.5 Corollary.</u> If M is real i.e. given by (1.7) with coefficients $a_n > 0$, $1/a_n$, a_0, $\ldots$, a_{n-1} locally integrable, then

(2.1) $$n \le d_+(M) = d_-(M) = d(M) \le 2n$$

and all values of d not excluded by (2.1) are realized.

   The lower bound in (2.1) is due to H. Weyl [130] for $n = 1$ and to I. M. Glassman [61] for general n. In the complex coefficient case the lower bound is due to Everitt [38]. Part (c) is due to Kogan and Rofe-Beketov [86] and was also established independently by Karlsson [78] in a more general setting and by Everitt. In the complex coefficient case constant coefficient expressions with unequal deficiency indices can easily be found in the odd order case. In the even order case with constant or non-constant complex coefficients examples with unequal deficiency indices are not easily found. The first such example is due to J. B. McLeod [96]. Kogan and Rofe-Beketov, in the paper mentioned above, developed a general method for getting such examples. For the proofs of the results of Theorem 2.3 and Corollary 2.4 the reader is referred to the papers mentioned above.

3. __Differential Operators__. Just as a comparison of a matrix with its transpose is useful in studying the dimension of the solution space of a linear algebraic system, so a comparison of a differential operator with its adjoint is useful in the study of the dimension of the space of square integrable solutions of a differential equation. For this purpose we associate differential operators in $L^2(a, \infty)$ with differential expressions and then study their adjoints in $L^2(a, \infty)$.

3.1 __Definition__. A differential expression with differentiable coefficients on an interval I is an expression of type

$$M = \sum_{j=0}^{m} p_j D^j$$

where $D = d/dt$ is the differentiation operator, p_j is a complex valued function with j continuous derivatives on I, and p_m is not identically zero on I. The order of M is m.

3.2 __Definition__. The expression M above is said to be regular on I for $I = [a, \infty)$ or I compact if p_m has no zero on I.

Below, $AC(I)$ denotes the set of absolutely continuous complex valued functions on I.

3.3 __Definition of maximal operator__. Let M be a regular differential expression with differentiable coefficients on I of order m. Then the maximal operator $T_1(M)$ associated with M is defined as follows: the domain of $T_1(M)$ is the set of all f in $L^2(I)$ such that $f^{(m-1)}$ is $AC(I)$ (so that Mf is defined a.e.) and Mf is in $L^2(0, \infty)$. Then $T_1(M)f$ is defined as Mf for all f in this domain.

3.4 __Definition__. Let M be a differential expression on I. Let C_0^∞ denote the set of infinitely differentiable functions each supported on a compact subset of the interior of I - so that each function in C_0^∞ vanishes in a neighborhood of both "end points" of I. Let $T_R(M)$ be the operator in $L^2(I)$ defined by $T_R(M)f = Mf$ for any f in C_0^∞.

3.5 __Definition of minimal operator__. Let M be a regular differential expression with differentiable coefficients on an interval I. The minimal operator $T_0(M)$ associated with M on I is defined as the smallest closed operator in $L^2(I)$ which extends $T_R(M)$. In other words, the domain of $T_0(M)$ consists of all f in $L^2(I)$ such that for some sequence $\{f_n\}$ in $C_0^\infty(I)$, $\{f_n\}$ converges to f and Mf_n converges to some g in $L^2(I)$. Then $T_0(M)f = g$.

As usual the inner product of f and g in $L^2(a, \infty)$ is denoted by (f, g).

3.6 Lemma. The minimal operator $T_0(M)$ is well defined.

Proof. Suppose there are two sequences $\{f_n\}$ and $\{g_n\}$ in C_0^∞, both converging to f in $L^2(I)$ with $\{Mf_n\}$ converging to u and $\{Mg_n\}$ converging to v in $L^2(I)$. We show that u = v.

From the Lagrange identity 2.1.1 we get $(M(f_n - g_n), \phi) = (f_n - g_n, M^+\phi)$ for any ϕ in $C_0^\infty(I)$. Hence $(M(f_n - g_n), \phi)$ converges to 0 for all such ϕ. On the other hand $\{(M(f_n - g_n), \phi)\}$ converges to $(u-v, \phi)$. Therefore $(u-v, \phi) = 0$ for any ϕ in C_0^∞ and consequently u = v.

Next we want to prove that $(T_0(M))^* = T_1(M^+)$, where $(T_0(M))^*$ denotes the adjoint operator (in $L^2(I)$) of $T_0(M)$. This will be done with the help of several lemmas. Some of these show that the operator $T_0(M)$ doesn't contain any "unreasonable" functions in its domain, others show that the domain of $(T_0(M))^*$ contains only "nice" differentiable functions. So that $T_0(M)$ and $(T_0(M))^*$ are differential operators.

An outline of the proof is as follows:
The fact that $(T_0(M))^* \supseteq T_1(M^+)$ follows easily from integration by parts. The reverse containment is more difficult. It is first shown for compact intervals and the general case deduced from this.

3.7 Lemma. Let M be a regular differential expression of order m with differentiable coefficients on an interval I containing its left end point a. Then $T_0(M) \subseteq T_1(M)$ and, for any f in the domain of $T_0(M)$, we have

$$f(a) = f'(a) = \ldots = f^{(m-1)}(a) = 0.$$

If I is compact, say I = [a, b], then also

$$f(b) = f'(b) = \ldots = f^{(m-1)}(b) = 0.$$

We remark that $T_1(M)$ contains $T_0(M)$ in general whether or not I contains its left end point, as we shall see later.

Proof. Let f_n be a Cauchy sequence of functions from $C_0^\infty(I)$ converging to f with Mf_n converging to T_0f in $L^2(I)$. It follows from the variation of parameters formula in the theory of ordinary linear differential equations that, on any compact interval $[a, b]$, $f_n^{(i)}$ converges uniformly, for all $i \le m-1$. Therefore $f_n^{(m)}$ also converges in $L^2(I)$. It follows that $f^{(m-1)}$ is absolutely continuous and $f^{(i)}(a) = 0$ for $1 \le i \le m-1$. If $I = [a, b]$, then also $f^{(i)}(b) = 0$ for $1 \le i \le m-1$. Since $f_n^{(i)}$ converges to $f^{(i)}$ for $i \le m$, we see that $T_0(M)f = Mf$, MF is in L^2 and so f is in the domain of $T_1(M)$.

3.8 Lemma. Let M be a regular differential expression with differentiable coefficients on I and order m. Suppose $T_1(M)f = 0$. Then $f^{(m)}$ is continuous on I.

Proof. Since $Mf = 0$ and $f^{(m-1)}$ is in AC, we have from

$$p_m f^{(m)} = - \sum_{j=0}^{m-1} p_j f^{(j)}$$

that $f^{(m)}$ is equal almost everywhere to a continuous function. Hence $f^{(m)}$ can be taken as continuous.

3.9 Lemma. Let M be a regular differential expression of order m with differentiable coefficients on a compact interval $I = [a, b]$. For f in $L^2(I)$, the equation $My = f$ has a solution y satisfying $y^{(i)}(a) = y^{(i)}(b) = 0$ for $0 \le i \le m-1$ if and only if f is orthogonal to the solution space of $M^+y = 0$.

Proof. Let $My = f$ and $M^+g = 0$. Then integration by parts shows that $(f, g) = (My, g) = 0$ if $y^{(i)}(a) = y^{(i)}(b) = 0$, $0 \le i \le m-1$.

Conversely, suppose that f is orthogonal to all solutions of $M^+y = 0$. Choose y such that $My = f$ and $y^{(i)}(a) = 0$, $0 \le i \le m$. We show that $y^{(i)}(b) = 0$ for $0 \le i \le m-1$. From the Lagrange identity 2.1.1 and with the notation used there we have $[y, z](b) - [y, z](a) = 0$ when $M^+z = 0$. Since $y^{(i)}(a) = 0$ for $0 \le i \le m-1$, it follows that $[y, z](b) = 0$ for all solutions z of $M^+z = 0$. For fixed i, $0 \leq i \leq m-1$, using the definition of $[y, z]$ and prescribing the values of $z^{(j)}(b)$, $0 \leq j \leq m-1$ appropriately we get $y^{(i)}(b) = 0$.

3.10 Lemma. Let S be a dense subspace of a Hilbert space H and suppose that V is a closed subspace of H with finite deficiency n. Then $V \cap S$ has deficiency n in S and is dense in V.

Proof. Clearly, any $n + 1$ nonzero vectors in H have a non-trivial linear combination in V. Thus the deficiency of $V \cap S$ in V is at most n. Suppose S_1 is a subspace of S such that $S \cap V + S_1 = S$. Then $V + S_1 \supseteq S$, so $V + S_1$ is dense in H. Since V is a closed subspace with finite deficiency in H and $V + S_1$ contains V, $V + S_1$ is a finite dimensional extension of V and is therefore closed. Thus $V + S_1 = H$. Hence S_1 is at least n dimensional.

We have shown that $V \cap S$ has deficiency n in S. If $V \cap S$ were not dense in V, then there would exist a vector $x \neq 0$ in V perpendicular to $V \cap S$. This would mean that there were $n + 1$ mutually orthogonal vectors x_i in H perpendicular to $V \cap S$. If S_1 is an n-dimensional subspace of S with $S_1 \cap V = \{0\}$ and $S_1 + V = S$, then it would follow that some linear combination of the $n + 1$ vectors x_i is orthogonal to S_1 and thus $V \cap S + S_1 = S$. This is a contradiction since S is dense in H.

3.11 Lemma. Let M be a regular differential expression of order m with differentiable coefficients on a compact interval $I = [a, b]$. Let A denote the restriction of $T_1(M)$ to the functions which vanish along with their first $m - 1$ derivatives at a and b. Then $A = T_0(M)$.

Remark. We have already shown that $T_0(M) \subseteq A$. It will follow from Lemma 3.9 that $A^* = T_1(M)$. Thus this lemma is crucial.

Proof. We first show the lemma for the case $M = D^m$. The general case follows from this. By Lemma 3.9, the range of A is the orthogonal complement in $L^2[a, b]$ of the m dimensional solution space of M^+. It follows from Lemma 3.10 that $C_0^\infty(I) \cap$ range A is dense in range A.

However, if $Af = g$ and g is in $C_0^\infty(I)$ then f vanishes in a neighborhood of each end point of I. Since g is infinitely differentiable and $M = D^m$ it follows that f is in $C_0^\infty(I)$. Let $Au = v$. Then there is a sequence g_n in $C_0^\infty(I) \cap$ range A with g_n converging to v in $L^2(I)$. Let $g_n = Af_n$. Then f_n is in $C_0^\infty(I)$ and it is clear, in addition, that for each $i < m$ $f_n^{(i)}$ converges to $u^{(i)}$. (The convergence is uniform for $i < m$). Thus $A = T_0(M)$ when $M = D^m$.

For a general M suppose that f is in the domain of A. Then $f^{(m-1)}$ is in $AC(I)$, $f^{(m)}$ is in $L^2(I)$ since $f^{(m)} = p_m^{-1}(Af - \sum_{i=0}^{m-1} p_i f^{(i)})$ and $f^{(i)}(a) = f^{(i)}(b) = 0$ for $1 \le i \le m-1$. Thus, by what we just proved, f is in the domain of $T_0(D^m)$. Furthermore there is a sequence f_n in C_0^∞ with $f_n^{(i)}$ converging to $f^{(i)}$ in $L^2(I)$ for all $i \le m$. Clearly Mf_n then converges to Mf so that f is in domain $T_0(M)$. This completes the proof.

3.12 **Lemma.** Suppose M is a regular differential expression with differentiable coefficients on a compact interval $I = [a, b]$. Then

$$(T_0(M))^* = T_1(M^+) .$$

Proof. It follows from Lemma 3.7 and integration by parts that $(T_0(M))^* \supseteq T_1(M^+)$.

Let f be in the domain of $(T_0(M))^*$. From the variation of parameters formula it follows that $T_1(M^+)$ is surjective since I is compact, so there is a vector g in domain $T_1(M^+)$ such that $T_1(M^+)g = T_0(M))^*f$. Thus $(T_0(M))^*(f-g) = 0$, so that if v is in the range of $T_0(M)$ with $Mu = v$, we have

$$(f-g, v) = (f-g, Mu) = ((T_0(M))^*(f-g), u) = 0.$$

Therefore $f - g$ is orthogonal to the range of $T_0(M) =$ range A. But, by Lemma 3.9, $f - g$ is then in the null space of $T_1(M^+)$. Hence f is in the domain $T_1(M^+)$ and $T_1(M^+)f = (T_0(M))^*f$. This completes the proof.

3.13 **Theorem.** Suppose M is a regular differential expression with differentiable coefficients on any interval I. Then

$$(T_0(M))^* = T_1(M^+).$$

Proof. Since $T_0(M)$ is the smallest closed extension of $T_R(M)$, it follows that $(T_0(M))* = (T_R(M))*$. Since, for any f in C_0^∞ and g in domain $T_1(M)$ we can see that

$$(T_R(M)f, g) = (f, T_1(M^+)g).$$

It follows that

$$T_1(M^+) \subseteq (T_R(M))* = (T_0(M))*.$$

If g is in domain $(T_0(M))*$, then on any compact subinterval [a, b] of I, we have $(Mf, g) = (f, (T_0(M))*g)$ for all f in $C_0^\infty(a, b)$. Thus the restriction of g to [a, b] is in the domain of R* where R is the restriction of $T_R(M)$ to $C_0^\infty(a, b)$. But by Lemma 3.8 we see that $g^{(m-1)}$ must be absolutely continuous on [a, b] and the restriction of $(T_0(M))*g$ to [a, b] must agree with M^+g on [a, b]. Since [a, b] is arbitrary, the theorem is proved.

3.14 Lemma. Let M be a regular differential expression with differentiable coefficients on any interval I. Then $(M^+)^+ = M$.

Proof. For all g in C_0^∞ and all f in C^m, where m = order M, we have

$$((M^+)^+f, g) = (Mf, g). \text{ Hence } (M^+)^+f = Mf \text{ and so } (M^+)^+ = M.$$

3.15 Corollary. For a differential expression M on any interval, $T_0(M) \subseteq T_1(M)$.

Proof. $T_1(M) = (T_0(M^+))*$ since $(M^+)^+ = M$. Thus $T_1(M)$ is a closed operator since all adjoint operators are closed. Since $T_1(M) \supseteq T_R(M)$ it follows that $T_1(M) \supseteq T_0(M)$.

3.16 Theorem. Let M be a regular differential expression with differentiable coefficients on an interval $[a, \infty)$. Then the range of $T_1(M)$ is dense in $L^2(a, \infty)$. Furthermore, the following statements are equivalent:

a) $T_0(M)$ has closed range

b) $T_1(M)$ is surjective

c) $T_0(M^+)$ has closed range.

Proof. The orthogonal complement of range $T_1(M)$ = null space $T_0(M^+)$ since $(T_1(M))* = T_0(M^+)$. Since $T_0(M^+)$ is one-to-one the first assertion of the theorem follows.

a → b. This is clear from the first part of the theorem since range $T_1(M)$ is a finite dimensional extension of range $T_0(M)$. (Range $T_0(M)$ = orthogonal complement of the null space of $(T_0(M))*$ = orthogonal complement of the null space of $T_1(M^+)$). So the range of $T_0(M)$ has finite deficiency in $L^2(a, \infty)$.

b → c. This follows from the general fact that if range A is closed then range A* is also closed.

c → a. If $T_0(M^+)$ has closed range, then $T_1(M^+)$ has closed range, so $T_0(M)$ has closed range.

4. General deficiency indices.

We have already defined the deficiency indices of a symmetric operator in chapter 1. We now define the deficiency index of a not necessarily symmetric differential operator and relate this definition to the earlier one.

4.1 Definition. Let M be a regular differential expression with differentiable coefficients. Define d(M), the mean deficiency index of M, to be one-half the dimension of the quotient space: domain $T_1(M)$/domain $T_0(M)$.

4.2 Theorem. Suppose that $T_0(M)$ has closed range, M being a regular expression, on $[a, \infty)$. Then $2d(M) = $ nullity $T_1(M^+) + $ nullity $T_1(M)$.

Proof. Note that (range $T_0(M))^\perp$ = null space $T_1(M^+)$. The operator $T_0(M)$ may be extended to $T_1(M)$ in two steps: first add to domain $T_0(M)$ a basis for the null space of $T_1(M)$ and then pick a basis $\{g_i\}$ for the null space of $T_1(M^+)$ and add to the domain $T_0(M)$ one f_i for each g_i such that $T_1(M)f_i = g_i$. This is possible since $T_1(M)$ is surjective by Theorem 3.16. We have constructed a basis for the quotient space domain $T_1(M)$/domain $T_0(M)$ and it has the proper number of elements, so the proof is complete.

4.3 Theorem. Let M be a regular differential expression with differentiable coefficients on $[a, \infty)$. Suppose $M = M^+$. Then $d(M) = [d_+(M) + d_-(M)]/2$.

Proof. This follows from von Neumann's formula 1.2.6, since it is clear that $T_0(M)$ is symmetric, being the closure of the symmetric operator $T_R(M)$.

4.4 Theorem. Suppose M is regular on $[a, \infty)$. Then

$$2d(M) \geq \text{nullity } T_1(M) + \text{nullity } T_1(M^+).$$

Remark. If the range of $T_0(M)$ is closed we have equality above; see Theorem 4.2.

Proof. Let $n = \text{nullity } T_1(M^+) = $ dimension of $(\text{range } T_0(M))^\perp$. Clearly, since $T_1(M)$ has dense range, at least n functions in domain $T_1(M)$ must be linearly independent modulo the subspace: null space $T_1(M) + $ domain $T_0(M)$. The conclusion follows from this.

4.5 Theorem. Let M be regular on $[a, \infty)$. Suppose f is a bounded continuous complex valued function on $[a, \infty)$. Then

$$d(M + f) = d(M).$$

Proof. This follows from the fact that domain $T_1(M) = $ domain $T_1(M + f)$ and domain $T_0(M) = $ domain $T_0(M + f)$. The first statement is obvious. The second follows from the fact that $T_0(M)$ is the closure of $T_R(M)$ and that if g_n converges to g and Mg_n converges to h, then $(M + f)g_n$ converges to $h + fg$.

4.6 Theorem. Let M be regular on $[a, \infty)$. Then $d(M) \geq (\text{order } M)/2$. Further, if, for some bounded continuous function f, $T_0(M + f)$ has closed range, then $d(M) \leq$ order M and $d(M) = d(M^+)$.

Remark. It is often possible to find a complex number λ such that $T_0(M + \lambda)$ has closed range. In particular this can be done when $M = M^+$ or when $T_0(M)$ is bounded below. When it is not possible to find an f as in the theorem, the determination of $d(M)$ becomes more complicated.

Proof. The second part of Theorem 4.6 follows from 4.5 and 4.2.

To prove the first part, note that any function in the domain of $T_0(M)$ vanishes along with its first $m - 1$ derivatives at a. Thus any set of functions $\{f_i\}$, $0 \leq i \leq m-1$ with the property that $f_i^{(j)} = \delta_{ij}$ – the Kronecker delta – is a linearly independent set modulo domain $T_0(M)$. Select such a set of f_i's where each f_i is in C^∞ and vanishes outside a compact interval. Then each such f_i is in the domain of $T_1(M)$ and the proof is complete.

4.7 Corollary. Suppose that M is as in Theorem 4.6 and $d(M) = $ (order M)$/2$. Then, if f is in domain $T_1(M)$ there is a g in domain $T_0(M)$ such that f - g is a C^∞ function supported in a compact interval.

Proof. If $d(M) = $ (order M)$/2$, the f_i's selected in the proof of Theorem 4.6 determine a basis for the quotient space: domain $T_1(M)/$domain $T_0(M)$. The Corollary follows from this.

4.8 Theorem. Let M be regular on $[a, \infty)$, $M = M^+$ and order $M = m$. Then

$$d_+(M) \geq [m/2] \quad \text{and} \quad d_-(M) \geq [m/2].$$

Proof. The minimal operator $T_0(M)$ is symmetric since it is the closure of a symmetric operator. Consider $m - [(m+1)/2]$ functions f_i with the following properties:

i) $f_i^{(j)}(a) = 0$, $0 \leq j \leq [(m+1)/2]$

ii) f_i in $C^\infty[a, \infty)$, and f_i supported in $[a, a + 1]$.

iii) $f_i^{(j+k)}(a) = \delta_{ik}$, $j = [(m+1)/2]$, $0 \leq k \leq m-j$.

These f_i are linearly independent modulo domain $T_0(M)$. Furthermore, if g and h are of the form

$$f + \sum_{i=1}^{j} \lambda_i f_i \ , \quad \text{with} \quad f \text{ in domain } T_0(M),$$

then $(Mg, h) = (g, Mh)$. Thus if A denotes the operator formed by restricting M to such functions g, then A is symmetric. Therefore $A + iI$ and $A - iI$ have no non-trivial null space. Now if, say, $d_-(M)$ were less than $[m/2]$ then it would follow that $S = $ null space of $T_1(M-iI)$ would be of dimension less then $[m/2]$, so that any $[m/2]$ functions in $L^2(a, \infty)$ would have a linear combination perpendicular to S. Since

$$S^\perp = \text{range} \ T_0(M + iI)$$

we would have

$$(M + iI) \sum \lambda_j f_j = T_0(M + iI)u. \quad \text{Then}$$
$$(A + iI)(\sum \lambda_j f_j - u) = 0,$$

a contradiction. The theorem is proved.

5. The general limit point condition.

5.1 **Definition.** Let M be a regular, not necessarily symmetric, differential expression with differentiable coefficients of order m on $[a, \infty)$. Then M is said to be in the limit point condition if $d(M) = m/2$.

Remark. Note that $d(M)$ need not be an integer. By Corollary 4.7, if M is in the limit point condition, any function in domain $T_1(M)$ agrees with a function in domain $T_0(M)$ on a semi-infinite interval. Thus the asymptotic behavior of solutions is determined by the domain of $T_0(M)$.

5.2 **Theorem.** Let M be regular on $[a, \infty)$. Then M is in the limit point condition if and only if

$$(5.1) \qquad \lim_{t \to \infty} [f, g](t) = 0$$

for all f in domain $T_1(M)$ and all g in domain $T_1(M^+)$ where $[f, g]$ denotes the Lagrange bilinear form of f and g.

Remark. Condition (5.1) is often used to derive sufficient conditions on the coefficients for the limit point case.

Proof. Suppose first that M is in the limit point condition. Then for any f in domain $T_1(M)$, $f(t) = h(t)$ for large t with h in domain $T_0(M)$. Thus in condition (5.1) we may take f in domain $T_0(M)$. For such an f and g in domain $T_1(M^+)$ we have

$$(Mf, g) = (f, M^+g). \quad \text{Thus}$$

$$\lim_{t \to \infty} [f, g] = -[f, g](a).$$

Now f in $T_0(M)$, we have $f^{(i)}(a) = 0$ for $i <$ order M. This implies $[f, g](a) = 0$ and the proof of the first half is complete.

Now suppose (5.1) holds. Let T denote the operator formed by restricting $T_1(M)$ to the set of all f such that $f^{(i)}(a) = 0$ for $i <$ order M. We show that $T = T_0(M)$ by showing that $T^* = T_1(M^+)$. If g is in domain T^*, then g is in domain $T_1(M^+)$ because $T \supseteq T_0(M)$. If g is in domain $T_1(M^+)$ it follows from the Lagrange identity and (5.1) that $(Mf, g) = (f, M^+g)$ for all f in domain T. Thus $T^* = T_1(M^+)$ and therefore, since T is clearly closed,

$$T = T** = T_0(M).$$

This means that, for any f in domain $T_1(M)$, $f = u + v$ with u in domain $T = T_0(M)$ and v a C^∞ compact support function such that $v^{(i)}(a) = f^{(i)}(a)$ for $i <$ order M. The set of such v's has dimension $m =$ order M modulo domain $T_0(M)$. Thus $d(M) = m$ and the theorem is proved.

5.3 Corollary. Let M be regular on $[a, \infty)$. Then M is in the limit point condition if and only if M^+ is in the limit point condition.

6. Notes and comments.

The smoothness assumptions $p_j \in C^j$ can be weakened substantially if the adjoint expression M^+, the Lagrange identity, etc. are expressed in terms of so-called quasi-derivatives. If an appropriate quasi-differential formulation is used, then only local integrability assumptions on the coefficients are needed. See Zettl [139]. It's interesting to note that (1.7) (or (1.6) in the complex case) represents the general form of symmetric expressions only when the coefficients are taken as very smooth. When the coefficients are allowed to be nondifferentiable there are much more general real symmetric quasi-differential expressions than (1.7) - see [139].

The coefficients p_j are assumed sufficiently differentiable in section 3 mainly for simplicity. With the given assumptions on p_j the domain of the minimal operator contains the C_0^∞ functions. Without smoothness assumptions this need not be the case and the description of the minimal domain is much more involved. Also it is then non-trivial to show that the minimal domain is dense.

The classification result (a) of Theorem 2.3 has an interesting history. It was established, in the real case, by H. Weyl in 1910 [130] for $n = 1$ and by I. M. Glassman in 1950 [61] for general n. In the 40 year interval there were several papers [131,116-120] published in which the authors claimed to have established that only the two cases $d(M) = n$ and $d(M) = 2n$ are possible. In chapter 5 we will give examples of operators M so that $d(M) = k$ for any k satisfying $n \leq k \leq 2n$. These examples are much simpler than Glassman's original ones.

The deficiency index problem for the whole line $(-\infty, \infty)$ can be reduced to the half-line case by means of the following formula (sometimes called Kodaira's formula):

$$d_+ = d'_+ + d''_+ - n; \quad d_- = d'_- + d''_- - n$$

where d_+, d'_+, d''_+ are the positive deficiency indices of an expression M of order n on the intervals $(-\infty, \infty)$, $(-\infty, 0]$, $[0, \infty)$, respectively. The indices d_-, d'_-, d''_- are defined similarly.

The notion of deficiency index for nonsymmetric expressions is due to Kauffman [79]. The proofs of the basic results on minimal and maximal operators in sections 3, 4 and 5 differ in some respects from the standard ones. See Dunford and Schwartz [21], Naimark [97] and Akhieser and Glassman [1] for other expositions.

At the International Conference on Differential Equations held in Uppsala, Sweden, during April, 1977, Gilbert announced the following result: Given any positive integer n there exists a symmetric expression M of type (1.6) such that

$$\left| d_+(M) - d_-(M) \right| \geq n.$$

This announcement will appear in the Proceedings of this Conference published by Springer-Verlag in the Lecture Notes series.

CHAPTER III

SECOND ORDER LIMIT-POINT, LIMIT-CIRCLE CONDITIONS

1. The general second order theory. This chapter contains an introduction to the classification theory of the symmetric second order expression

(1.1) $My = -(py')' + qy$

on the interval $[a, \infty)$. A comprehensive treatment of this much studied subject is outside the scope of these Notes. Instead we shall restrict ourselves to a selection of criteria with relatively simple proofs which illustrate the different types of results which have been obtained, with some emphasis on those relevant to the discussion of the deficiency indices of powers of differential expressions in chapter 6. For reference purposes a more complete list of results, without proofs, appears in the Notes and Comments section at the end of this chapter.

Except when otherwise indicated we shall assume throughout this chapter that p and q are real-valued functions on $[a, \infty)$ with p positive, that p is C^2 and q is continuous. These smoothness assumptions are made so that the general theory as developed in chapter 2 is available here and could be replaced in most cases by the assumptions that p^{-1} and q are Lebesque integrable on each finite interval $[a, x]$.

By 2.2.3 and 2.2.5 the positive and negative deficiency indices always coincide and are equal to either 1 or 2. When $d(M) = d_+(M) = d_-(M) = 1$ we say that we are in the limit-point case. When $d(M) = 2$ we speak of the limit-circle case. These terms were originally introduced for (1.1) and refer to geometric features of the construction, based upon complex function theory, by which it was originally established by Hermann Weyl that these are the only possibilities. For an account of this approach see the original classic paper of Weyl [130] or Titchmarsh [121] or Coddington-Levinson.

In order to classify (1.1) it suffices to examine the solutions of $My = 0$ because of the following simple fact.

1.1 Lemma. If for some complex number λ_0 all solutions of $My = \lambda_0 y$ are in

$L^2(a, \infty)$, then all solutions of $My = \lambda y$ are in $L^2(a, \infty)$ for every complex number λ.

Remark. Thus M is limit-point if and only if there is a solution of $My = 0$ which is not square integrable.

Proof. Let u and v be linearly independent solutions of $My = \lambda_0 y$ for which

$$p(uv' - u'v) = 1.$$

Let λ be an arbitrary complex number and let z be any solution of $My = \lambda y$ or, equivalently, $(M - \lambda_0)z = (\lambda - \lambda_0)z$. Then by the variation of parameters formula there are constants b and c such that

$$(1.2) \qquad z(t) = bu(t) + cv(t) + (\lambda - \lambda_0) \int_a^t [u(t)v(s) - u(s)v(t)]z(s)\,ds.$$

Set $||f||^2(t) = \int_a^t |f|^2$ and $||f|| = \sup ||f||(t)$ when this is finite. Then by the Schwarz inequality applied to $\int uz$ and $\int vz$ the integral in (1.2) is less than

$$[|u|(t) \, ||v||(t) + |v(t)| \, ||u||(t)] \, ||z||(t)$$

$$\leq (||u|| + ||v||)(|u|(t) + |v|(t)) \, ||z||(t).$$

Hence from (1.2),

$$||z||^2(t) \leq 4(|b|^2 + |c|^2)(||u|| + ||v||)^2 +$$

$$4|\lambda - \lambda_0|^2(||u|| + ||v||)^2 \int_a^t (|u| + |v|)^2(s) \, ||z||^2(s)\,ds.$$

By Gronwall's inequality,

$$||z||^2(t) \leq K \exp \int_a^t (|u| + |v|)^2$$

$$\leq K \exp 2(||u||^2 + ||v||^2)$$

for all t. Hence z is in $L^2(a, \infty)$ and the proof is complete.

2. Limit-point criteria. Perhaps the simplest limit point criterion is the following one.

2.1 Theorem. For arbitrary p, if q is non-negative, then $My = 0$ has a positive increasing solution and thus is limit-point.

Proof. Let z be the solution of (1.1) with $z'(a) = 1$, $z(a) = 0$. Let b be the smallest zero of z in (a, ∞). Then z' has a zero c in (a, b). But

$$(2.1) \qquad p(c)z'(c) = p(a) + \int_a^c qz \geq p(a) > 0.$$

Hence z and z' are both positive on (a, ∞).

2.2 Corollary. For arbitrary p, if q is bounded below, then M is limit-point.

Proof. This is a consequence of Theorem 2.1 and Theorem 2.4.5.

2.3 Theorem. For arbitrary p, if q is in $L^2(a, \infty)$, then M is limit-point.

Proof. Suppose $Mz = 0$ and z is in $L^2(a, \infty)$. Then qz is in $L^1(a, \infty)$ and pz' has a finite limit as $t \to \infty$ by (2.1). Thus if u and v are solutions of $My = 0$ with $p(u'v - uv') = 1$ which are in $L^2(a, \infty)$, then

$$1 \leq K(|u| + |v|).$$

But this is impossible since the right hand side is in $L^2(a, \infty)$.

2.4 Corollary. For arbitrary p, if q is in $L^r(a, \infty)$ with $r \geq 2$, then M is limit-point.

Proof. Set $q = q_1 + q_2$ where $q_1(t) = q(t)$ when $|q(t)| \leq 1$ and $q_1(t) = 1$ when $q(t) > 1$ and $q_1(t) = -1$ when $q(t) < -1$. Then q_1 is bounded and $q_2 = q - q_1$ is in $L^2(a, \infty)$. Hence M is limit-point by Theorems 2.3 and 2.4.5.

The next criterion is not quite as elementary. It is designed to investigate when M is limit-point even though q has large negative values on a substantial subset of $[a, \infty)$. We shall see in various specializations of this result that in order for M to be limit-point it suffices that q be well behaved on only a part of its domain and, further, that the size of this part is related to the behavior of q there.

2.5 Theorem. Suppose that there is a non-negative locally absolutely continuous function w and a decomposition $q = q_1 + q_2$ of q such that

   i) $(1 + \delta)pw'^2 - q_1w^2 \leq K$ for some $\delta > 0$

   ii) $wp^{-\frac{1}{2}}|Q| \leq K$ for some function Q with $Q' = q_2$

   iii) $\int_a^\infty wp^{-\frac{1}{2}} = \infty$.

Then M is limit-point.

2.6 Remark. The function q_2 may be thought of as a perturbation of q_1 which is restricted in terms of p and q_1 by (ii). If q_1 is negative, then each term on the left in (i) is bounded by K and w may be thought of as approximately $(-q)^{-\frac{1}{2}}$. For sufficiently nice expressions M and negative q the size of the solutions of $My = 0$ is approximately $(-pq)^{-\frac{1}{4}}$. (see Remark 3.3 and Corollary 3.5). From this point of view it is (iii) that really asserts that $My = 0$ has a solution which is not square-integrable.

Proof. It will be shown that py'^2w^2 is in $L^1(a, \infty)$. Assuming this, suppose that u and v are solutions with $p(u'v - uv') = 1$. Then

$$(2.2) \qquad (p^{\frac{1}{2}}u'w)v - (p^{\frac{1}{2}}v'w)u = wp^{-\frac{1}{2}}$$

so that if both u and v are in $L^2(a, \infty)$, then the left hand side of (2.2) is in $L^1(a, \infty)$ by Schwarz's inequality. But the right hand side of (2.2) is not in $L^1(a, \infty)$ by (iii). Thus u and v cannot both be in $L^2(a, \infty)$. Hence M is limit-point.

To establish the assertion concerning py'^2w^2, suppose that $My = 0$ with y in $L^2(a, \infty)$. Then

$$- \int_a^t (py')'yw^2 + \int_a^t (q_1 + q_2)y^2w^2 = 0$$

or, integrating by parts,

$$(2.3) \qquad \int_a^t py'^2w^2 = -2\int_a^t py'yw'w - \int_a^t (q_1 + q_2)y^2w^2 + py'yw^2 \Big|_a^t .$$

Let $\delta > 0$ be as in (i) and choose $\theta < 1$ so that $(1 + \delta)\theta > 1$. Set $\varepsilon_1 = (1 + \delta)(1 - \theta)$ and choose $\varepsilon_2 > 0$ so that $[(1 + \delta)\theta]^{-1} + \varepsilon_2 < 1$. Using the elementary inequality $2ab \le \varepsilon^{-1}a^2 + \varepsilon b^2$, for any $\varepsilon > 0$, we have

$$(2.4) \qquad -2 \int_a^t py'yw'w \le [(1 + \delta)\theta]^{-1} \int_a^t py'^2w^2 + (1 + \delta)\theta \int_a^t pw'^2 .$$

Also, integrating by parts again,

$$- \int_a^t q_2y^2w^2 = 2 \int_a^t Qy^2w'w + 2 \int_a^t Qy'yw^2 - Qy^2w^2 \Big|_a^t .$$

The first integral on the right here satisfies, using (ii),

$$(2.5) \qquad 2 \int_a^t Qy^2w'w \le 2K \int_a^t p^{\frac{1}{2}}w'y^2$$

$$\le \varepsilon_1 \int_a^t pw'^2y^2 + (K/\varepsilon_1) \int_a^t y^2 .$$

Similarly, for the second integral we have

$$(2.6) \qquad 2 \int_a^t Qy'yw^2 \le 2K \int_a^t p^{\frac{1}{2}}y'yw$$

$$\le \varepsilon_2 \int_a^t py'^2w^2 + (K^2/\varepsilon_2) \int_a^t y^2 .$$

Inserting (2.4), (2.5), and (2.6) into (2.3), we obtain

$$(1 - [(1 + \delta)\theta]^{-1} - \varepsilon_1) \int_a^t py'^2 w^2$$

$$\leq (1 + \delta) \int_a^t pw'^2 y^2 - \int_a^t q_1 w^2 y^2 + K_1 \int_a^t y^2$$

(2.7)

$$+ \; py'yw^2 - Qy^2w^2 \; \Big|_a^t$$

$$\leq (K_1 + K_2) \int_a^t y^2 + py'yw^2 - Qy^2w^2 \; \Big|_a^t$$

by (i).

Now by (ii),

$$py'yw^2 - Qy^2w^2 \leq py'yw^2 + Kp^{\frac{1}{2}}y^2w.$$

We shall show that this function cannot be eventually bounded away from 0. This, together with the last inequality and (2.7), is enough to show that py'^2w^2 is in $L^1(0, \infty)$ as asserted.

Thus, suppose $py'yw^2 + Kp^{\frac{1}{2}}y^2w \geq c > 0$ on some interval $[t_0, \infty)$. Set

$$N = \{t \geq t_0 : \; y'(t)y(t) < 0\}.$$

Then on N, $Kp^{\frac{1}{2}}y^2w > |py'yw^2|$ so that $Kp^{-\frac{1}{2}}w^{-1} > |y'/y|$. Also on N, $Kp^{\frac{1}{2}}y^2w \geq c$ so $(K^2/c)y^2 \geq Kp^{-\frac{1}{2}}w^{-1}$. Thus

$$- \int_N y'/y < (K^2/c) \int_N y^2 < \infty.$$

But then $\log|y|$ is bounded below on $[t_0, \infty)$ so that y is bounded away from 0 there. This is impossible since y is in $L^2(a, \infty)$. Thus py'^2w^2 is in $L^1(a, \infty)$ and the proof is complete.

It is now possible to get some insight into when M can be limit-point despite q being unbounded below.

2.6 Example. Let $My = -y'' - t^\alpha y$, $t \geq 1$. In Theorem 2.5 take $q = q_1$ and $w(t) = t^{-1}$. Then the hypotheses of Theorem 2.5 are satisfied if and only if $\alpha \leq 2$. It will be shown in section 3 that M is limit-circle when $\alpha > 2$.

In fact the same choice for w applies whenever $-q(t) \leq Kt^2$ for some K and all t. One natural generalization of this inequality is as follows.

2.7 Corollary. If there is a positive locally absolutely continuous function W with $-q \leq W$, $W'W^{-3/2}$ essentially bounded and $\int_a^\infty W^{-\frac{1}{2}} = \infty$, then $-y'' + qy$ is limit-point.

Proof. In Theorem 2.5 take $q = q_1$, $w = W^{-\frac{1}{2}}$.

These examples appear to establish a limit on the permissable rate of growth of q. However $-y'' + qy$ can be limit-point even though this limit is repeatedly and dramatically violated by q.

2.8 Example. Set $My = -y'' - (t^2 + te^t \cos(e^t))y$, $t \geq 1$. Here Theorem 2.5 applies with $w(t) = t^{-1}$, $q_1(t) = -t^2$, $q_2(t) = -te^t \cos(e^t)$.

It is also possible for $-q(t) \leq Kt^2$ to fail most of the time even with $q \leq 0$.

2.9 Example. Set $My = -y'' - t^3 \sin^4 t$, $t \geq 0$. Here $-q(t) \geq t^{5/2}$ except on a sequence of intervals about $n\pi$ whose lengths approach zero. However its behavior on these intervals is sufficient to make M limit-point. To see this, set $q = q_1$ and on each interval $I_n = [n\pi - n^{-\frac{1}{2}}, n\pi + n^{-\frac{1}{2}}]$ define w by

$$w(t) = \begin{cases} t - n\pi + n^{-\frac{1}{2}}, & n\pi - n^{-\frac{1}{2}} \leq t \leq n\pi, \\ w(2n\pi - t), & n\pi \leq t \leq n\pi + n^{-\frac{1}{2}}. \end{cases}$$

Set $w = 0$ on the complement of the union of the I_n's. For each n, $\int_{I_n} w = 1/n$. On I_n, $|w'(t)| \leq 1$, $w(t) \leq n^{-\frac{1}{2}}$, and $-q(t) \leq n$. Thus the hypotheses of Theorem 2.5 are satisfied.

By refining the techniques of the previous two examples we can obtain the following generalization of Example 2.6 and the remark following it. We shall write q^+ and q^- for the positive and negative parts of q.

2.10 Corollary. If $\int_a^t q^- \leq Kt^3$ for all $t \geq b$, then $-y'' + qy$ is limit-point. Proof. The technique will be to regard $[a, \infty)$ as the union of a sequence of subintervals and on each of these to decompose q^- into the sum of two functions — its integral mean value on the subinterval and the difference between q_1 and this function. The second of these has integral zero over each subinterval and will serve as q_2. The first is comparatively small pointwise, at least on sufficiently many subintervals, and its sum with q^+ will become q_1.

Choose a positive integer N so that $2^N \geq a$. For each $n \geq N$ set $I_n = [2^n, 2^{n+1}]$ and divide each such interval into 2^{2n} subintervals, each of length 2^{-n}. Since $\int_{I_n} q^- \le 8K2^{3n}$, it follows that on at least half of these

subintervals, J,

(2.8) $\qquad \int_J q^- \le 16K2^n$.

(Otherwise the sum obtained by adding the integrals of q^- over the subintervals

would exceed $8K2^{3n}$.) On each subinterval $J = [c, d]$ for which (2.8) holds,

define w by

$$w(t) = \begin{cases} t - c, & c \le t \le c + 2^{-n-1} \\ w(c + d - t), & c + 2^{-n-1} \le t \le d. \end{cases}$$

Then $\max_J w = 2^{-n-1}$ and $\int_J w = 2^{-2n-2}$. On each subinterval where (2.8) does

not hold and on $[a, b]$, set $w = 0$. Then for each $n \ge N$, $\int_{I_n} w \ge 2^{2n-1} \cdot 2^{-2n-2} =$

$1/8$. Hence $\int_a^\infty w = \infty$.

Next define a step function q_0 which is constant on $[a, b]$ and on each

subinterval (whether (2.8) holds or not) and is equal on each of these subintervals

to the mean value of q^- there. Thus on each $[c, d]$

$$q_0 = \int_c^d q^- / (d - c).$$

On a subinterval where (2.8) holds,

$$q_0 w^2 \le (\int_c^d q^-) 2^n (2^{-n-1})^2 \le 4K.$$

On subintervals where (2.8) fails, and on $[a, b]$, $q_0 w^2 = 0$.

Now decompose q by setting $q_1 = q^+ - q_0$, $q_2 = q_0 - q^-$. Then $w'^2 \le 1$ and

$-q_1 w^2 \le 4K$ so (i) in Theorem 2.5 holds. Also $w(t) \int_a^t q_2$ is different from 0 only

when t lies in a subinterval $J = [c, d]$ for which (2.8) holds. For such t,

by the construction of q_0,

$$w(t) \int_a^t q_2 = w(t) \int_c^t q_2 \le \max_J w \int_J q^- \le 8K,$$

so that (ii) also holds. Finally, (iii) has already been verified so the Corollary

now follows from Theorem 2.5.

We consider now to what extent these results carry over to the general second

order expression

$$My = -(py')' + qy.$$

For non-constant p it is roughly the case that the more rapidly p increases the more rigid are the restrictions on the rate of growth of $-q$.

2.11 Example. Set $My = -(t^\alpha y')' - t^\beta y$, $t \geq 1$, $\alpha \leq 2$. With $w(t) = t^{\alpha/2 - 1}$ and $q = q_1$ the hypotheses of Theorem 2.5 are satisfied if $\beta \leq 2 - \alpha$. As with Example 2.6, it will be shown in section 3 that M is limit-circle for larger values of β. Curiously, as will be shown in section 4, M is limit-point for some positive values of β when $\alpha > 2$.

A version of Corollary 2.7 for general p follows from the same choice of w as in that result.

2.12 Corollary. If there is a positive locally absolutely continuous function W with $-q \leq W$, $pW'^2 W^{-3}$ essentially bounded, and $\int_a^\infty (pW)^{-\frac{1}{2}} = \infty$, then M is limit-point.

Note that $W(t) = (\int_a^t p^{-\frac{1}{2}})^2 (= (t - a)^2$ when $p = 1)$ satisfies the hypotheses of Corollary 2.12.

One striking feature of Example 2.9 and Corollary 2.10 is that the limit-point classification is completely independent of the behavior of q on the rather large set where $w = 0$. We conclude this section with a result in which information about the coefficients of M is required only on a disjoint sequence of intervals which may be chosen completely arbitrarily, provided of course only that the intervals go to infinity and the coefficients are suitable there. By requiring that q be positive on the intervals we shall be able to make full use of (i) of Theorem 2.5.

2.13 Corollary. Suppose that there is a sequence $I_m = [a_m, b_m]$ of intervals on which $q(t) \geq q_m > 0$. Set $P_m = \int_{a_m}^{b_m} p^{-\frac{1}{2}}$.

If for some $c < 1/2$,

$$\sum_{m=1}^\infty [\exp(cq_m^{\frac{1}{2}} P_m) - 1]P_m^2 = \infty,$$

then M is limit-point.

Proof. Fix $d < 1$ with $d^2 > 2c$. Then for some positive constant K,

(2.9) $[\exp(d^2 x/2) - 1]/dx > K(\exp(cx) - 1)$

for all $x > 0$.

Decompose each I_m into a union

$$I_m = \cup_{j=0}^{3} [c_j, c_{j+1}]$$

of four intervals with common endpoints so that $a_m = c_0$, $b_m = c_4$ and

$$\int_{c_1}^{c_2} p^{-\frac{1}{2}} = \int_{c_2}^{c_3} p^{-\frac{1}{2}} = dP_m/2; \quad \int_{c_0}^{c_1} p^{-\frac{1}{2}} = \int_{c_3}^{c_4} p^{-\frac{1}{2}} = (1 - d)P_m/2.$$

Set $w = 0$ outside the union of the I_m's. On I_m take $q_1 = q$ and define w as follows. On $[c_0, c_1]$: $w(t) = \int_{c_0}^{t} p^{-\frac{1}{2}}$. Then $w(c_1) = (1 - d)P_m/2$ and $pw'^2 = 1$ on (c_0, c_1).

On $[c_1, c_2]$: $w(t) = w(c_1) \exp d \int_{c_1}^{t} (q_m/p)^{\frac{1}{2}}$. Then $w' = d(q_m/p)^{\frac{1}{2}} w$ or

$$d^{-2}pw'^2 - qw^2 \le d^{-2}pw'^2 - q_m w^2 = 0$$

on (c_1, c_2).

On $[c_2, c_3]$: $w(t) = w(c_1) \exp d \int_{t}^{c_3} (q_m/p)^{-\frac{1}{2}}$. Then again

$$d^{-2}pw'^2 - qw^2 \le d^{-2}pw'^2 - q_m w^2 = 0$$

on $[c_2, c_3]$.

On $[c_3, c_4]$: $w(t) = \int_{t}^{c_4} p^{-\frac{1}{2}}$.

Then $pw'^2 = 1$ on (c_3, c_4) and $w(c_4) = 0$.

Thus hypotheses (i) and (ii) of Theorem 2.5 are satisfied with this choice of w. It remains only to verify (iii). For each m,

$$\int_{a_m}^{b_m} wp^{-\frac{1}{2}} > \int_{c_1}^{c_2} wp^{-\frac{1}{2}}$$

$$= w(c_1)(dq_m^{\frac{1}{2}})^{-1}[\exp d \int_{c_1}^{c_2}(q_m/p)^{\frac{1}{2}} - 1]$$

$$= [(1 - d)/2]P_m^2[\exp d^2 q_m^{\frac{1}{2}} P_m - 1]/dq_m^{\frac{1}{2}} P_m$$

$$> ((1 - d)K/2)[\exp cq_m^{\frac{1}{2}} P_m - 1]P_m^2$$

by (2.9). Since K is independent of m it now follows that

$$\int_{a}^{\infty} wp^{-\frac{1}{2}} = \infty$$

and hence that this result is a consequence of Theorem 2.5.

2.14 Example. Set $My = -(t^\alpha y')' + t^\beta \sin(t^\gamma)y$, $t \ge 1$, $0 \le \alpha \le 2$, $\beta \ge 0$, $\gamma \ge 0$. With $I_m = [(2m\pi + \pi/6)^{1/\gamma}, (2m\pi + 5\pi/6)^{1/\gamma}]$, $q_m^{\frac{1}{2}} P_m$ behaves like m^c with $c = (\beta + 2 - \alpha - 2\gamma)/2\gamma$. Thus M is limit-point provided $\beta - \alpha > 2\gamma - 2$. On the other hand it follows directly from Theorem 2.5 with $w(t) = t^{-1 + \alpha/2}$ and $q = q_2$ that M is limit-point if $\beta \le \gamma$. Surprisingly, however, for $\alpha = 0$

it has been shown [3] by asymptotic methods that M is limit-circle for the "wedge" $\gamma < \beta < 2\gamma - 2$ not covered by either inequality.

The fact that Corollary 2.13 does not restrict the lengths of the intervals I_m in any way has the following remarkable consequence.

2.15 Corollary. If M is any limit-circle expression, then for any $\varepsilon > 0$ there is a limit-point expression whose coefficients coincide with those of M on the complement of a set whose Lebesgue measure is less than ε.

3. The limit-circle case. It follows from Corollary 2.15 that limit-circle criteria must be of a rather different type from those of Theorem 2.5. We shall give two results which are really comparison principles. The first one will be required in section 4 with noncontinuous coefficients.

3.1 Theorem. Set $M_j y = -(py')' + q_j y$, $j = 1, 2$ where p is positive and p, q_1, and q_2 are Lebesque integrable on each finite interval $[a, x]$. If $|q_1 - q_2| y^2$ is in $L^1(0, \infty)$ whenever $M_1 y = 0$, then every solution of $M_1 y = 0$ is in $L^2(0, \infty)$ if and only if every solution of $M_2 y = 0$ is in $L^2(0, \infty)$.

Proof. Let u and v be solutions of $M_1 y = 0$ for which $p(u'v - uv') = 1$. Set $r^2 = u^2 + v^2$. If $M_2 z = 0$, then by the variation of parameters formula there are constants b and c such that

$$z(t) = bu(t) + cv(t) + \int_a^t [u(t)v(s) - u(s)v(t)][q_1(s) - q_2(s)]z(s)ds.$$

Hence

$$|z(t)| \leq (|b| + |c|)r(t) + r(t) \int_a^t r|q_1 - q_2| |z|,$$

or with $x = z/r$

$$|x(t)| \leq (|b| + |c|) + \int_a^t r^2 |q_1 - q_2| |x|.$$

Hence by Gronwall's inequality,

$$|x| \leq (|b| + |c|)\exp \int_a^\infty r^2|q_1 - q_2| \leq K$$

so that $|z| \leq Kr$.

Thus certainly all solutions of $M_2 y = 0$ are in $L^2(0, \infty)$ if all solutions of $M_1 y = 0$ are in $L^2(0, \infty)$.

For the converse, the calculation shows that if $M_2 z = 0$, then

$$|q_1 - q_2|z^2 \leq K^2|q_1 - q_2|r^2.$$

Thus the hypothesis also holds for solutions of $M_2 y = 0$ and the roles of M_1 and M_2 may be reversed.

We now return to the assumption that p is C^2 and q is continuous.

3.2 Theorem. If there is a positive C^2 function f such that $f[Mf + 1/pf^3]$ is in $L^1(0, \infty)$, then M is limit-circle if and only if f is in $L^2(0, \infty)$.

Proof. Set $M_2 = M$ and $M_1 y = -(py')' + [(pf')'/f - (pf^4)^{-1}]y$. Then the solutions of $M_1 y = 0$ are

$$y(t) = f(t)\sin(c + \int_a^t (pf^2)^{-1})$$

for arbitrary c. For such y,

$$|q_1 - q_2|y^2 \leq f|qf - (pf')' + (pf^3)^{-1}|$$
$$= f|Mf + (pf^3)^{-1}|.$$

Thus by Theorem 3.1, $M = M_2$ is limit-circle if and only if M_1 is limit-circle, i.e. if and only if f is in $L^2(0, \infty)$.

3.3 Remark. If $Mu = 0 = Mv$ and $r^2 = u^2 + v^2$ then it is readily verified that

$$Mr + 1/pr^3 = 0.$$

By definition, M is limit-circle if and only if r is in L^2. Thus f may be considered to be a "sufficiently good approximation" to r.

3.4 Example. Set $My = -(t^\alpha y')' - t^\beta y$, $t \geq 1$. Set $f(t) = t^{-(\alpha+\beta)/4}$. Then

$$f[Mf + 1/pf^3] = Kt^{-2 + (\alpha-\beta)/2}$$

which is in $L^1(0, \infty)$ if $\alpha - \beta < 2$. Thus, for such α and β, M is limit-circle if and only if $\beta > 2 - \alpha$. In particular, $-y'' - t^\beta y$ is limit-point when $\beta \leq 2$ and limit-circle when $\beta > 2$.

A natural generalization of this example provides a result related to the choice $W = (-q)^{\frac{1}{2}}$ in Corollary 2.12.

3.5 Corollary. Suppose that q is negative and C^2. Set $f = (-pq)^{-\frac{1}{4}}$. If

$f(pf')'$ is in $L^1(0, \infty)$, then M is limit-circle if and only if f is in $L^2(0, \infty)$.

3.6 Remark. Theorem 3.2 can apply even when q is not as smooth as this on the entire half-line and, in particular, can be used to show that a limit-circle expression can be perturbed on a sufficiently small set without affecting the limit circle property. We illustrate this for the special case $\alpha = 0$ of Example 3.4.

Let $\beta > 2$ and let E be any unbounded subset of $[0, \infty)$ for which

$$\int_E t^{\beta/2} = \infty .$$

Define $q(t) = -t^{\beta}$ on the complement of E and $f(t) = t^{-\beta/4}$ for all nonnegative t. Then the hypotheses of Theorem 3.2 will be satisfied for any choice of q on E for which

$$\int_E |q|t^{-\beta/2} < \infty.$$

This will certainly be the case if $|q(t)| \leq t^{\beta}$ on E. Thus for any such q, M will still be limit-circle. On the other hand, by Corollary 2.15 it is possible to change q on E so that M becomes limit-point.

3.7 Remark. The device in the proof of Theorem 3.2 may also be used to show that for any positive C^2 function p there is a continuous function q such that $My = -(py')' + qy$ is limit-circle. Let f be any positive C^2 function which is in $L^2(0, \infty)$. Set $q = (pf')' - (pf^4)^{-1}$. Then every solution of $My = 0$ is of the form

$$f(t)\sin(c + \int_a^t (pf^2)^{-1})$$

and so is in $L^2(0, \infty)$.

4. Oscillation and the limit-point classification. This final section is concerned with the relationship between oscillation theory and the limit-point, limit-circle theory. We start by developing some properties and examples of non-oscillatory expressions and then develop some consequences of these for the limit-point classification of these expressions.

4.1 Example. By Theorem 2.1, $My = -(py')' + qy$ is non-oscillatory if q is nonnegative (and $p > 0$).

It is also possible for M to be nonoscillatory when q is negative.

4.2 Example. The expression $-(t^\alpha y')' - (\frac{1}{4})t^{\alpha-2}y$, $t \geq 1$ can be seen to be nonoscillatory for $\alpha \leq 0$ and $\alpha \geq 2$ by actually computing the solutions. They are powers of t.

Some nonoscillatory expressions with oscillatory q will be given following Proposition 4.8.

4.3 Proposition. If M is nonoscillatory, then for any nontrivial solutions x and z of $My = 0$, x/z is eventually monotonic. There is a solution v, unique up to constant multiples, such that

$$\lim_{t\to\infty} v(t)/z(t) = 0$$

for each solution z linearly independent of v. Moreover,

$$(4.1) \qquad \int^\infty (pv^2)^{-1} = \infty; \quad \int^\infty (pz^2)^{-1} < \infty$$

if z is not a multiple of v.

4.4 Definition. The solution v is called the principal solution. All solutions linearly independent of v are nonprincipal.

Proof. For nontrivial x and z,

$$(x/z)' = W(x, z)/z^2 = c(pz^2)^{-1}$$

is defined and has constant sign beyond the largest zero of z. Hence $L = \lim x/z$ as $t \to \infty$ exists in the extended real numbers. For any linearly independent x and z, set $v = x - Lz$ if L is finite and $v = z$ otherwise. Then the first assertion concerning v is clear. The assertions about the integrals follow from

$$c \int_a^\infty (pz^2)^{-1} + x(a)/z(a) = \lim_{t\to\infty} x/z.$$

4.5. Proposition. The equation $My = 0$ has a positive solution x on an interval

$[a, b] = I$ if and only if the Riccati equation

$$(4.2) \qquad w' + w^2/p = q$$

has a solution on all of I.

Proof. Given a positive solution x, $w = px'/x$ satisfies (4.2) on I. Conversely, if w satisfies (4.2) and $x(t) = \exp \int_a^t w/p$ then $Mx = 0$ on I.

4.6 Remark. If (4.2) has a solution on $[a, b)$ which is not continuable to b and if p and q are continuous at b, then the corresponding solution x must have a zero at b. Hence $\lim \inf w(t) = -\infty$ as $t \to b^-$.

Solutions of Riccati equations have the following very useful comparison property.

4.7 Proposition. If $q_+ \geq q$ on an interval $I = [a, b]$, if $z' + z^2/p = q$ on I and if $w_0 \geq z(a)$, then the solution w_+ of the initial value problem $w' + w^2/p = q_+$, $w(a) = w_0$ exists on all of I and satisfies $w_+ \geq z$ there.

Proof. First suppose $w_0 = z(a)$. For $\varepsilon > 0$ let w_ε satisfy

$$(4.3) \qquad w_\varepsilon' + w_\varepsilon^2/p = q_+ + \varepsilon, \quad w_\varepsilon(a) = z(a).$$

Then w_ε exists on some subinterval $[a, c)$ of I. Since $w_\varepsilon(a) = z(a)$ and $w_\varepsilon'(a) = q_+(a) + \varepsilon \geq z'(a) + \varepsilon$, it follows that $w_\varepsilon > z$ on a (possibly smaller) subinterval $[a, c']$. Set $d = \inf\{t > a \mid w_\varepsilon(t) < z(t)\}$. Then $w_\varepsilon'(d) \leq z(d)$. But $w_\varepsilon(d) = z(d)$ so that $w_\varepsilon'(d) \geq z'(d) + \varepsilon$ from (4.3). Thus it must be the case that $w_\varepsilon \geq z$ on domain w_ε. By Remark 4.6, w_ε is defined on I and satisfies $w_\varepsilon \geq z$ there. Since $w_\varepsilon \to w_+$ uniformly as $\varepsilon \to 0$, w_+ is defined on I and $w_+ \geq z$ there. If $w_0 > z(a)$ then the result follows from Remark 4.6 again and the uniqueness of solutions of (4.2).

4.8 Proposition. The equation $My = 0$ has a positive solution on $[a, \infty)$ if and only if $q = q_1 + q_2$ with

$$(4.4) \qquad \left(c + \int_a^t q_2\right)^2 \leq p(t)q_1(t) \quad \text{for all} \quad t \geq a \quad \text{and some} \quad c.$$

Proof. Suppose x is a positive solution of $My = 0$. Set $w = px'/x$. Then $w' + w^2/p = q$ and the condition is satisfied with $q_1 = w^2/p$ and $q_2 = w'$.

On the other hand, suppose that $q = q_1 + q_2$ and (4.4) holds. Set $Q(t) = c + \int_a^t q_2$ and $z = py' - Qy$. Then $My = 0$ can be expressed as a system:

$$y' = (Q/p)y + (1/p)z$$
$$z' = (q_1 - Q^2/p)y - (Q/p)z.$$

Let $w = (z/y)\exp \int_a^t Q/p$. Then

$$(4.5) \qquad w' + w^2/(p \exp \int_a^t Q/p) = (q_1 - Q^2/p) \exp \int_a^t Q/p \geq 0.$$

It follows from Proposition 4.7 with $z = q = 0$ and the right hand side of (4.5) as q_+ that (4.5) has a non-negative solution w on $[a, \infty)$. Then

$$y(t) = \exp \int_a^t [Q/p + w/(p \exp \int_a^s Q/p)]ds$$

is a positive solution of $My = 0$.

We can now easily give examples of nonoscillatory expressions with oscillatory coefficients.

4.9 Example. It follows from Proposition 4.8 with $q_1 = 1$ and $q_2 = e^t \cos e^t$ that $-y'' + (1 + e^t \cos e^t)y$ is nonoscillatory on $[0, \infty)$.

4.10 Example. Similarly, $-(t^\alpha y')' + (t^\beta + t^\gamma \cos(t^\delta))y$ is nonoscillatory on $[1, \infty)$ provided $\alpha + \beta > \max\{0, 2(\gamma + \delta + 1)\}$.

We consider now the limit-point classification of nonoscillatory expressions.

4.11 Theorem. If M is nonoscillatory and $p^{-\frac{1}{2}}$ is not in $L^1(0, \infty)$, then M is limit-point.

Proof. Proof Let u be a nonprincipal solution of $My = 0$. By Schwarz's inequality,

$$\infty = \left(\int_a^\infty p^{-\frac{1}{2}}\right)^2 \leq \int_a^\infty (pu^2)^{-1} \int_a^\infty u^2 .$$

The first term on the right is finite by Proposition 4.3.

4.12 Remark. Note that Example 4.9 is limit-point and Example 4.10 is limit-point when $\alpha \leq 2$.

4.13 Remark. The restriction on p in Theorem 4.11 is best possible in the sense that for any p with $p^{-\frac{1}{2}}$ in $L^1(0, \infty)$, there is a function q such that

$My = -(py')' + qy$ is nonoscillatory and limit-circle. To see this, just set $z = p^{-\frac{1}{4}}$ and $q = (pz')'/z$. Then $Mz = 0$ and z is in $L^2(0, \infty)$. Moreover, z is a nonprincipal solution of $My = 0$ since $(pz^2)^{-1} = p^{-\frac{1}{2}}$ is in $L^1(0, \infty)$. Thus every solution of $My = 0$ is in $L^2(0, \infty)$.

There is also a comparison theorem for nonoscillatory expressions.

4.14 Theorem. If $q_2 \geq q_1$ on $[a, \infty)$ and if $M_1 y = -(py')' + q_1 y$ is nonoscillatory and limit-point, then $M_2 y = -(py')' + q_2 y$ is limit-point.

Proof. Let u be an eventually positive nonprincipal solution of $M_1 y = 0$. Then u is not square-integrable. Choose b so that $u(t) > 0$ for $t \geq b$, and set $z = pu'/u$ there. Then $z' + z^2/p = q_1$ on $[b, \infty)$. By Proposition 4.7 there is a function w on $[b, \infty)$ such that $w(b) = z(b)$, $w' + w^2/p = q_2$ and $w \geq z$. Thus

$$x(t) = u(b) \exp \int_b^t w/p \geq u(b) \exp \int_b^t z/p = u(t)$$

is a solution of $M_2 y = 0$ on $[b, \infty)$ which is not square-integrable there. Clearly x can be extended to a solution of $M_2 y = 0$ on $[a, \infty)$. Hence M_2 is limit-point.

4.15 Remark. A remarkable point about Theorem 4.13 is that it is false without the hypothesis that M_1 is nonoscillatory. In fact, given any continuous function q on $[0, \infty)$ there is a continuous $r \leq q$ such that $-y'' + ry$ is limit-point. To see this, set $q_n = \min\{q(t) \mid n \leq t \leq n+1\}$. Let $\{k_n\}$ be a sequence of even positive integers such that $(\pi k_n)^2 \geq q_n$. Define a step function r_0 by

$$r_0(t) = -(\pi k_n)^2, \quad n \leq t < n+1.$$

Then $-y'' + r_0 y = 0$ has a solution u given by

$$u(t) = \cos(\pi k_n(t - n)), \quad n \leq t \leq n+1.$$

Note that $\int_n^{n+1} u^2 = 1/2$ for each n so u is not in $L^2(0, \infty)$. All solutions of this equation are bounded since another solution is

$$v(t) = (\pi k_n)^{-1} \sin(\pi k_n(t - n)), \quad n \leq t \leq n+1.$$

Hence if $r < r_0$ and is continuous and $r_0 - r$ is in $L^1(0, \infty)$, then it follows from Theorem 3.1 that $-y'' + ry$ is limit-point.

4.16 Remark. We may also use Theorem 4.14 to complete the discussion of

(4.6) $-(t^\alpha y')' - t^\beta y, \quad t \geq 1.$

Recall from Example 3.4 that if $\alpha - \beta < 2$, then (4.6) is limit-point if $\beta \leq 2 - \alpha$ and limit-circle if $\beta > 2 - \alpha$. Also by Theorem 2.4.5 and Theorem 2.1 we have that (4.6) is certainly limit-point when $\beta \leq 0$ for any α. It remains to consider the wedge $0 < \beta \leq \alpha - 2$, $\alpha > 2$. It is easy to see that

(4.7) $-(t^\alpha y')' - Kt^{\alpha-2}y = 0$

has a solution which is a power of t and is not square integrable if and only if $K \leq (2\alpha - 3)/4$. In particular, (4.6) is limit-point for $\beta = \alpha - 2 > 0$ if and only if $\alpha \geq 7/2$. Moreover, by comparing (4.6) to a suitable version of (4.7) and using Theorem 4.14, we see that (4.6) is always limit-point when $\beta < \alpha - 2$. Note that the limit-point classification of (4.7) for arbitrary positive K is the same as that of (4.6) except on the ray $\beta = \alpha - 2$, $\alpha > 2$ where the point at which the expression changes from limit-circle to limit-point is $\alpha = (4K + 3)/2$.

5. Notes and comments.

Corollary 2.2 dates from the initial work on the deficiency index problem by Hermann Weyl [125] in 1910. Theorem 2.2 is due to Putnam [105] and is one of a number of results obtained during the late 1940's in the great resurgence of interest in differential equations inspired by Aurel Wintner and Philip Hartman.

Theorem 2.5 is one of a series of results of a roughly similar type which includes work by Hartman and Wintner [69], Levinson [95], and, more recently, Atkinson and Evans [4]. Levinson's theorem appears here as Corollary 2.12. The following related result, due to Sears [115], is also a consequence of Theorem 2.5. Theorem. If there is a positive monotonic function W with $-q \leq W$ and $\int_a^\infty (pW)^{-\frac{1}{2}} = \infty$, then $My = -(py')' + qy$ is limit-point.

The first condition generally like (ii) of Theorem 2.5 was introduced by Brinck [7] for a slightly different purpose. His result was adapted as a limit-point criterion of the "Levinson type" in a greatly strengthened form by Knowles [85].

A form of Theorem 2.5 in which q may be complex-valued appears in Read [112]. Atkinson and Evans [4] and Atkinson [2] also permit complex-valued q.

Corollary 2.10 is a theorem of Hartman and Wintner [68]. Their proof depends on an investigation of the distribution of zeros of a solution of $-y'' + qy = 0$. This method was refined by Eastham [22] to yield a criterion in which estimates of $\int q^-$ are required only on a sequence of disjoint intervals. This result, and a related result of Atkinson [2, Theorem 11] can be derived from Theorem 2.5 in much the same way as Corollary 2.10. See Read [112] for details.

Another "interval criterion" has recently been established by Evans [32]. It is as follows.

<u>Theorem.</u> Suppose that in each of a sequence $I_m = [a_m, b_m]$ of disjoint intervals there exist a real locally intebrable function Q_m, a non-negative function k_m, a positive absolutely continuous function W_m and positive constants δ, K, G_m such that with $q = q_1 - q_2$, $(q_1, q_2$ locally integrable), we have in I_m,

A) $q_1(x) \geq (1 + \delta)H_m^2/p(x) - k_m(x)$, $H_m = \int Q_m$,

B) $\int_\alpha^\beta (q_2 - Q_m)W_m \leq KG_m p_m^{\frac{1}{2}}$ for all $a_m \leq \alpha < \beta \leq b_m$

where $p_m = \inf\{p(t): t \in I_m\}$.

Suppose also that there exists a real piecewise continuously differentiable function ϕ_m with support in I_m such that

C) $\sum_{m=1}^\infty \phi_m^{-1} \int_{I_m} \frac{W_m^2}{(q_1 + k_m)^{\frac{1}{2}}} p^{-\frac{1}{2}} Q_m^2 = \infty$

where $\phi_m = \sup_{I_m} \{(G_m^2 + p_m^{\frac{1}{2}} G_m|W_m'| + k_m W_m^2)\phi_m^2 + p(W_m\phi_m)'^2\}$. Then $-(py')' + qy$ is limit-point.

The hypothesis $f(pf')'$ in L_1 made in Corollary 3.5 actually suffices for the existence of an asymptotic formula for the solutions of $-(py')' + qy = 0$ from which the conclusion also follows (see, for instance, Coppel [11]). However the argument given here is simpler and suffices for our purposes.

Section 4. Propositions 4.3, 4.5, and 4.7 are standard results in the theory of disconjugacy. See, for instance, Coppel [10]. Proposition 4.8 is essentially Theorem 6.1 of Read [111].

Theorem 4.11 and Remark 4.13 are due to Hartman [69]. Theorem 4.14 appears in Kurss [91]. The construction in Remark 4.15 is a generalization by Eastham and Thompson [27] of an example of Hartman and Wintner [68].

The first interval criterion requiring q to be positive but not restricting the lengths of the intervals was established by Ismagilov [75] for p = 1 and by Knowles [84] for general p. A slightly stronger result than Corollary 2.13 which requires no pointwise estimates of q (except q ≥ 0) is given in Read [114].

Example 2.14 was investigated by Atkinson, Eastham and McLeod [3]. Corollary 2.15 is suggested by a closely related result of Eastham and Thompson [27] (see also the notes for section 4.)

Section 3

Theorem 3.1 is due to Halvorsen [65] and Theorem 3.2 to Knowles [83]. The following more general limit-circle criterion has been proved by Kupcov [90].

__Theorem.__ The expression $-(py')' + qy$ is limit-circle if there exist positive functions Q and ψ with Q' and ψ locally absolutely continuous and

$$\int_a^\infty \psi^2(t) \exp \int_a^t [|\frac{\psi'}{\psi} - \frac{Q'}{Q}| + \psi^2| - \frac{(pQ')'}{Q} + q - \frac{1}{p\psi^4} |]ds \, dt < \infty.$$

It should be noted from the proof of Theorem 3.1 that Theorem 3.2 could be reformulated as the special case of this result with $\psi = Q$ with no change in proof. However the form given suffices for most examples.

CHAPTER IV

HIGHER ORDER LIMIT-POINT CRITERIA.

1. Introduction.

In this chapter we establish some general limit-point conditions for $2n^{th}$ order real symmetric ordinary differential expressions. These conditions include the well known criterion of Hinton [67] and also yield strong "interval" type limit-point conditions. Our proof is different from that of Hinton and is taken from [37].

Consider real symmetric $2n^{th}$ order differential expressions

$$(1.1) \qquad My = \sum_{j=0}^{n} (-1)^{j}(p_{n-j}y^{(j)})^{(j)}$$

on the interval $[0, \infty)$. The coefficients are assumed real and satisfy the basic conditions

$$(1.2) \qquad p_0 > 0 \quad \text{and} \quad 1/p_0, \ p_{n-j} \in L^1_{loc} \ , \quad j = 0,1,\ldots,n-1.$$

2. The main result and some implications.

For convenience we introduce quasi-derivatives $y^{[j]}$ as follows:

$$(1.3) \qquad y^{[j]} = y^{(j)}, \quad j = 0,\ldots,n-1, \quad y^{[n]} = p_0 y^{(n)},$$
$$y^{[n+j]} = -(y^{[n+j-1]})' + p_{n-j}y^{(n-j)}, \quad j = 1,\ldots,n.$$

Whenever we write Mf we shall assume without further mention that f has locally absolutely continous quasi-derivatives up to order $2n - 1$ so that Mf exists a.e. and is locally integrable.

2.1 Theorem. Take $n > 1$. Let R be a non-negative differentiable function on $[0, \infty)$ and suppose that R^{4n}/p_0 and R^{4n-2}/p_0 have absolutely continous $(n-1)$th order derivatives on $[0, \infty)$. Let

(i) $p_i = r_i + q_i, \quad i = 1,2,\ldots,n$

and let h_i be such that

(ii) $q_i = h_i^{(i)}, \quad i = 1,2,\ldots,n-1$

(iii) $q_n \geq h_n^{(n)}.$

Suppose there exist positive constants K_1, K_2, ..., K_7, δ and a, such that for $t \geq a$

    (iv) $|r_i| R^{4i}/P_0 \leq K_1$, $i = 1, ..., n-1$,

    (v) $R^{2n-2i} |h_{n-i}|/P_0 \leq K_2$, $i = 0, 1, ..., n-1$,

    (vi) $(1 + \delta)[P_0(R^{4n}/P_0)^{(n)}]^2 - 4R^{8n}r_n/P_0 \leq K_3 R^{4n}$,

    (vii) $|RR'| \leq K_5$,

  (viii) $|(R^{4n}/P_0)^{(j)}| \leq K_6(R^{4n-2j}/P_0)$, $j = 1, 2, ..., n-1$,

    (ix) $|(R^{4n-2}/P_0)^{(j)}| \leq K_7 R^{4n-2-2j}/P_0$, $j = 1, 2, ..., n-1$

    (x) $\int_a^\infty R^{4n-2}/P_0 = \infty$.

Then M is limit-point.

Remarks.

1. We will see in chapter VII that the conclusion of Theorem 2.1 can be strengthened to read: Then $p(M)$ is limit-point for any polynomial p with real coefficients.

2. The special case when R is strictly positive and the two terms on the left in (vi) are required to be separately bounded by the right includes Hinton's conditions. In addition, Hinton assumed that $|R^2 P_0'/P_0| \leq K$. This last condition is the obstacle that prevented Hinton's result from reducing to that of Levinson when $n = 1$.

3. Allowing R to be non-negative rather than positive is significant in that it gives rise to so-called "interval type" criteria. Some of these will be mentioned below.

    Taking $R(t) = t^{(\alpha-1)/(4n-2)}$ we get

2.2 Corollary. Let $P_0(t) = t^\alpha$, $\alpha \leq 2n$, $|p_i| = O(t^{\gamma_i})$ where $\gamma_i = [4i + \alpha(4n - 4i - 2)]/(4n - 2)$, $i = 1, 2, ..., n-1$ and $p_n(t) \geq -Kt^c$ with $c = (4n - 2\alpha)/(4n - 2)$ for some $K > 0$, $t \geq a$. Then M is limit-point.

    Applying this Corollary to the special case

$$My = y^{(4)} + qy$$

we have that M is limit-point if

$$q(t) \geq -Kt^{4/3} \text{ for some } K > 0.$$

The power $4/3$ is known to be best possible here [40].

2.3 Corollary. Let $P_0 = 1$ and suppose there exists a sequence of disjoint intervals I_m of length μ_m such that in each I_m

    i) $P_i = h_{i,m}^{(i)}$, $i = 1,2,\ldots,n-1$

    ii) $P_n \geq h_{n,m}^{(n)}$

    iii) $\mu_m^{n-i} |h_{n-i,m}| \leq K$, $i = 0,1,\ldots,n-1$

    iv) $\sum_{m=1}^{\infty} \mu_m^{2n} = \infty$.

Then M is limit-point.

As a special case of Corollary 2.3 we have that M is limit-point if $P_0 = 1$ and there exists a sequence of intervals I_m of length μ_m with $\mu_m \geq \delta > 0$ for all m such that on these intervals $|P_i| \leq K_i$, $i = 1,2,\ldots,n-1$ and $P_n \geq -K_n$.

Proof. This follows easily from Theorem 2.1 on taking $r_i = 0$, $i = 1,\ldots,n$ and

$$R(t) = \begin{cases} (r-a_m)^{\frac{1}{2}} , & a_m \leq t \leq \frac{1}{2}(a_m + b_m) \\ (b_m-t)^{\frac{1}{2}} , & \frac{1}{2}(a_m + b_m) \leq t \leq b_m . \end{cases}$$

Now we apply some of the above results to the example:

(2.1) $My = y^{(2n)} + t^{\alpha}\sin t^{\beta} y$.

2.4 Corollary. If M is given by (2.6) then M^k , $k = 1,2,\ldots$ is limit-point under any of the following conditions:

    i) $\alpha \leq 2n/(2n-1)$,   all β

    ii) $\beta \leq 2n/(2n-1)$,   all $\alpha \geq 0$

    iii) $\alpha \leq n\beta - 2n(n-1)/(2n-1)$.

Proof.   i) Let $p_n(t) = r_n(t) = t^{\alpha}\sin t^{\beta}$ and $R(t) = t^{-1/(4n-2)}$ in Theorem 2.1

         ii) In Corollary 2.3 let I_m be given by

               $2m\pi + \delta \leq t^{\beta} \leq (2n+1)\pi - \delta$, $0 < \delta < \pi/2$.

Then $\mu_m \sim ((\pi-2\delta)(2\pi)^{1/\beta-1} \beta-1)m^{1/\beta-1}$

and $\sum_{m=1}^{\infty} \mu_m^{2n} = \infty$ if $\beta \leq 2n/(2n-1)$.

Choose $h_{n,m}(t) = t^{n(1-1/\beta)}$ so that $p_n \geq h_{n,m}^{(n)}$ on I_m and

$$\mu_m^n |h_{n,m}| = 0(1) \ .$$

iii) In Theorem 2.1 we choose $r_n = 0$ and

(2.2)
$$h_n^{(n)}(t) = t^\alpha \sin t^\beta \ , \quad R(t) = t^{-1/(4n-2)} \ .$$

The only condition that needs checking is (iv) in Theorem 2.1 with $i = 0$.

Let

$$I_{k,\ell}(t) = \int t^{\alpha-k\beta+\ell-1} \sin(t^\beta - k\pi/2)$$
$$= \beta^{-1} \int t^{\alpha-(k+1)\beta+\ell} [\sin\{t^\beta-(k+1)\pi/2\}]^{(1)} \ .$$

Then, on integrating by parts

(2.3)
$$I_{k,\ell}(t) = \beta^{-1} t^{\alpha-(k+1)\beta+\ell} \sin(t^\beta-(k+1)\pi/2) - \beta^{-1}(\alpha-(k+1)\beta+\ell) I_{k+1,\ell}(t)$$

and this gives, for some constants $C(m,k,\ell)$

(2.4)
$$I_{k,\ell}(t) = \sum_{m=k+1}^{n} C(m,k,\ell) \, t^{\alpha-m\beta+\ell} \sin(t^\beta-m\pi/2) + 0(t^{\alpha-n\beta+\ell}) \ .$$

From (2.2) and (2.3) it follows that

$$h_n^{(n-j)}(t) = \sum_{k_1=1}^{n} \sum_{k_2=k_1+1}^{n} \cdots \sum_{k_j=k_{j-1}}^{n} C(k_1,\ldots,k_j) t^{\alpha-k_j\beta+j} \sin(t^\beta-k_j\pi/2)$$
$$+ 0(t^{\alpha-n\beta+j})$$

and hence

(2.5)
$$h_n(t) = 0(t^{\alpha-n\beta+n}) \ .$$

Thus, for $\alpha \leq n\beta - 2n(n-1)/2n-1$,

$$R^{2n}|h_n(t)| \leq K$$

as required.

In the case $n = 1$ it was shown in Example 3.2.14 (with β in place of α and γ in place of β) that M is limit-point also when

(2.6)
$$\alpha > 2\beta - 2 \ .$$

In [3] a complete classification of (2.1) was given when $n = 1$. For $\beta < \alpha < 2\beta - 2$ $(\beta > 2)$ M is limit-circle. When $\alpha = 2\beta - 2$ both cases are possible.

Question. Let $n = 2$. For what values of α, β is $d(M) = 3$ for M in (4.5)? Are there any values of α, β such that $d(M) = 4$, i.e. M is limit-circle? An affirmative answer to the last question would be particularly interesting as there are presently no known examples of functions q such that $y^{(2n)} + qy$ is limit-circle for $n > 1$.

Remark 4. When $n = 1$, it was shown in chapter III that given _any_ sequence of intervals I_m going to infinity and any limit-circle expression M it is possible, by altering the coefficients p_0, p_1 only on the intervals I_m, to make M limit-point.

We know from chapter II that $d(M)$, the deficiency index of M, can take on any value n, $n+1$, ..., $2n$. From Corollary 4.1 we see that given any expression M, no matter what the value of $d(M)$, and any sequence of intervals I_m of length r_m subject only to the condition $\sum_{m=1}^{\infty} r_m^{2n} = \infty$, by changing the coefficients of M only on the intervals I_m we can arrange for $d(M) = n$. This indicates that the cases $d(M) > n$ are strikingly unstable under "small" perturbations. In particular, for any of the cases $d(M) > n$ it is not possible to get strong interval type sufficient conditions - strong in the sense that (iv) of Corollary 4.5 cannot be satisfied by the complements of the intervals. It is interesting to note that the very existence of these powerful interval criteria above for the limit-point case precludes the possibility of obtaining strong interval results for any other case.

Another indication of the delicate dependence of $d(M)$ on the coefficients was given by Kwong [97] who showed that $d(M)$ is not invariant under $L^1(0,\infty)$ perturbations of the coefficient p_n. This contrasts with the well known invariance of $d(M)$ under a bounded perturbation of p_n.

3. Proof of Theorem 2.1. The proof is lengthy, somewhat technical and is established with the help of a few lemmas.

The integrals appearing below are with respect to Lebesque measure, the dx symbol is omitted for brevity. The symbols K, K_1, K_2, ... will denote various positive constants and ε, ε_1, ε_2, ... will denote various "small" positive constants. These constants will not necessarily be the same on each occurrence. Also, we write $K(\varepsilon)$ when we wish to indicate the dependence of K on ε. Throughout this chapter the functions considered are real valued unless the contrary is explicitly stated.

3.1 Lemma. Let $n > 1$. Suppose there is a function R satisfying the conditions of Theorem 2.1. Then there exists a function R_1 and a positive number a_1 such that R_1 satisfies the conditions on R in Theorem 2.1 for $t \geq a_1$ and, in addition, R_1^{4n}/p_0 is bounded for $t \geq a_1$.

Proof. Let

$$R_1(t) = R(t)/\{\theta(t)\}^{1/(4n-2)}$$

where

$$\theta(t) = \int_a^t \{R^{4n-2}/p_0\} \; .$$

Choose a_1 such that $\theta(t) \geq 1$ for $t \geq a_1$. This is possible in view of (x) in Theorem 2.1. Hence $R_1 \leq R$ on $[a_1, \infty)$ and so (iv) and (v) hold. Also

$$\int_{a_1}^T (R_1^{4n-2}/p_0) = \log \theta(T) - \log \theta(a_1) \to \infty$$

as $T \to \infty$ so that (x) holds.

Now we show that R_1^{4n}/p_0 is bounded. The rest of the lemma will follow from this. Let $t_0 > a_1$. If $(R^{4n}/p_0)(t_0) \leq 1$, then $(R_1^{4n}/p_0)(t_0) \leq 1$. On the other hand, if $(R^{4n}/p_0)(t) > 1$ put $s_0 = \max\{t \,|\, a_1 \leq t < t_0, \; (R^{4n}/p_0)(t) \leq 1\}$ with $s_0 = a_1$ if the set exhibited is empty. Then, from (viii),

$$(R^{4n}/p_0)(t_0) = (R^{4n}/p_0)(s_0) + \int_{s_0}^{t_0} (R^{4n}/p_0)' \leq K(1 + \theta(t_0))$$

and hence $(R_1^{4n}/p_0)(t_0) \leq K$.

Thus, R_1^{4n}/p_0 is bounded for $t \geq a_1$. Condition (vii) follows immediately. For (viii) and (ix) we first show that

(3.1) $$\theta(1/\theta)^{(s)} R_1^{2s} \leq K_1, \quad s = 1,2,\ldots,n, \quad t \geq a_1,$$

(3.2) $$\theta^{4n/(4n-2)} (1/\theta^{4n/(4n-2)}) R^{2s} \leq K_2, \quad s = 1,2,\ldots,n, \quad t \geq a_1.$$

Both are proved by induction. For the first we have

$$-\theta(1/\theta)' = R^{4n-2}/(p_0\theta) = R_1^{4n-2}/p_0 \ .$$

Multiplying by R_1^2 gives (3.1) for $s = 1$.

Suppose (3.1) holds for all $s \leq j$. From Leibnitz's formula for the derivatives of products

$$R_1^{2s}(1/\theta^2)^{(s)} = R_1^{2s} \sum_{k=0}^{s} \binom{s}{k}(1/\theta)^{(s-k)}(1/\theta)^{(k)}$$

$$= 0(1/\theta^2)$$

and so

$$-R_1^{2(j+1)}\theta(1/\theta)^{(j+1)} = R_1^{2j+2}\theta(\theta'/\theta^2)^{(j)}$$

$$= \theta \sum_{s=0}^{j}\binom{j}{s}R_1^{2j+2-2s}\theta^{(j+1-s)}R_1^{2s}(1/\theta^2)^{(s)}$$

$$= \theta \sum_{s=0}^{j}\binom{j}{s}R_1^{2j+2-2s}(R^{4n-2-2j+2s}/p_0)(1/\theta)$$

$$= 0 \sum_{s=0}^{j}\binom{j}{s}R^{2(j+1-s)} \ 2(j+1-s)/(4n-2)(R^{4n-2(j+1-s)}$$

$$/p_0)\theta^{-1}$$

$$= 0(R^{4n}/p_0) = 0(1)$$

where we have used (ix) and $\theta(t) \geq 1$ for $t \geq a_1$. Hence (3.1) is established and from this it is easily verified that R_1 satisfies (ix) for $t \geq a_1$. Similar arguments can be used to prove (3.2) from which (viii) follows for $t \geq a_1$.

The next lemma will also be used in chapter VII when the limit-point case for powers of differential expressions is discussed.

3.2 Lemma. Let R be as in Theorem 2.1 and let v be a non-negative function with support in a compact interval $I \subset (0,\infty)$ and with an absolutely continuous $(n-1)$th derivative on I. Suppose that there exist positive constants K_i, $i = 1,2,\ldots,9$ independent of I such that i)-ix) in Theorem 2.1 are satisfied on I and also

a) $|R^{4n}/p_0| \leq K_4$

b) $|p_0 v^{(j)}| \leq K_8 \ R^{4n-2j}$, $j = 1,2,\ldots,n$

c) $|v| \leq K_9$.

Let $g = Mf$ and, for some real number ,

$$F_j = F_j(\alpha,f) = \int_I v^{2j+\alpha} \ R^{4j}(f^{(j)})^2 \ , \quad j = 0,1,\ldots,n$$

$$G_j = G_j(\alpha,f) = \int v^{2j+4n+\alpha} \ R^{4j}(g^{(j)})^2 \ , \quad j = 0,1,\ldots,n \ .$$

(3.4)

Then, given any $\varepsilon > 0$, there exists a positive constant $K(\varepsilon)$, independent of I, such that

(3.5)
$$F_j \leq \varepsilon F_{j+1} + K(\varepsilon)F_{j-1} \; , \quad j = 1,2,\ldots,n-1 \quad (n > 1)$$

(3.6)
$$F_n \leq \varepsilon G_0 + K(\varepsilon) \sum_{j=0}^{n-1} F_j$$

(3.7)
$$G_j \leq \varepsilon G_{j+1} + K(\varepsilon)G_{j-1} \; , \quad j = 1,2,\ldots,n-1 \quad (n > 1)$$

(3.8)
$$G_n \leq \varepsilon \int v^{8n+\alpha} (M^2 f)^2 + K(\varepsilon) \sum_{j=0}^{n-1} G_j \; .$$

Proof. The proof involves the use of integration by parts, Leibnitz's formula for derivatives of a product, and the simple inequality

$$2|ab| \leq \varepsilon a^2 + (1/\varepsilon)b^2$$

which holds for arbitrary $\varepsilon > 0$. In the integration by parts the constants of integration are all zero since v vanishes at the end points of I. All integrals are over I.

To establish (3.5)

$$F_j = \int v^{2j+\alpha} R^{4j} (f^{(j)})^2$$

$$= -\int v^{2j+\alpha} R^{4j} f^{(j+1)} f^{(j-1)}$$

$$\quad - \int \{(2j+\alpha)v^{2j-1+\alpha} v' R^{4j} + 4j v^{2j+\alpha} R^{4j-1} R'\} f^{(j)} f^{(j-1)}$$

$$\leq (F_{j+1} F_{j-1})^{\frac{1}{2}} + K \int v^{2j-1+\alpha} R^{4j-2} |f^{(j)} f^{(j-1)}|$$

on using the Schwartz inequality in the first integral and (a), (b), (c) and (vii) in the second after writing $|v'R^{4j}| = |v'p_0 R^{4j}/p_0|$

$$\leq K_1 R^{4n-2} R^{4j}/p_0$$

$$= K_1 R^{4n}/p_0 \, R^{4j-2}$$

$$\leq K_2 \, R^{4j-2}$$

$$\leq (F_{j+1} F_{j-1})^{\frac{1}{2}} + K(F_j F_{j-1})^{\frac{1}{2}}$$

$$\leq \varepsilon_1 F_{j+1} + K(\varepsilon_1)F_{j-1} + (\tfrac{1}{2})F_j + KF_{j-1} \; .$$

Thus (3.5) follows. Inequality (3.7) is established similarly. For (3.6) we have

$$F_n = \int v^{2n+\alpha} R^{4n} (f^{(n)})^2$$

$$= (-1)^n \int \{(v^{2n+\alpha} R^{4n}/p_0)(p_0 f^{(n)})\}^{(n)} f$$

$$= \sum_{\ell=0}^n (-1)^n \binom{n}{\ell} \int \{(v^{2n+\alpha} R^{4n}/p_0)^{(\ell)} (p_0 f^{(n)})^{(n-\ell)}\} f$$

by Leibnitz's formula,

$$= \int v^{2n+\alpha} (R^{4n}/p_0) f\{Mf - \sum_{j=0}^{n-1} (-1)^j (p_{n-j} f^{(j)})^{(j)}\}$$

$$+ \sum_{\ell=1}^n (-1)^n \binom{n}{\ell} \int \{(v^{2n+\alpha} R^{4n}/p_0)^{(\ell)} (p_0 f^{(n)})^{(n-\ell)}\} f$$

$$= \int \{v^{2n+\alpha} (R^{4n}/p_0) fMf\}$$

$$- \sum_{j=0}^{n-1} \int \{(v^{2n+\alpha}(R^{4n}/p_0) f)^{(j)} p_{n-j} f^{(j)}\}$$

$$+ \sum_{\ell=1}^n (-1)^\ell \binom{n}{\ell} \int \{(v^{2n+\alpha} R^{4n}/p_0)^{(\ell)} f\}^{(n-\ell)} p_0 f^{(n)}$$

on integration by parts,

$$= \int v^{2n+\alpha} (R^{4n}/p_0) fMf$$

$$- \sum_{j=1}^{n-1} \sum_{s=0}^j \binom{j}{s} \int (v^{2n+\alpha} R^{4n}/p_0)^{(j-s)} f^{(s)} p_{n-j} f^{(j)}$$

$$+ \sum_{\ell=1}^n \sum_{m=1}^{n-\ell} (-1)^\ell \binom{n}{\ell} \binom{n-\ell}{m} \int (v^{2n+\alpha} R^{4n}/p_0)^{(n-m)}$$

$$f^{(m)} p_0 f^{(n)}$$

$$- \int (v^{2n+\alpha} R^{4n}/p_0) p_n f^2$$

$$+ \sum_{\ell=1}^n (-1)^\ell \binom{n}{\ell} \int (v^{2n+\alpha} R^{4n}/p_0)^{(n)} p_0 ff^{(n)}$$

(3.9) $$= I_1 + I_2 + I_3 + I_4 + I_5$$

say. We shall need the estimates

(3.10) $$(v^{2n+\alpha})^{(\ell)} = O(v^{2n+\alpha-\ell} R^{4n-2\ell}/p_0), \quad \ell = 1, \ldots, n$$

(3.11) $$(v^{2n+\alpha} R^{4n}/p_0)^{(j)} = O(v^{2n+\alpha-j} R^{4n-2j}/p_0), \quad j = 0, 1, \ldots, n-1.$$

These are derived as follows: From (ix), (a) and (c) we have for $s \geq 0$, $t \geq 0$, $s + t \geq 1$, and $1 \leq k$, $j \leq n$,

$$(v^{(k)})^s (v^{(j)})^t = O(R^{4n(s+t)-2(sk+jt)}/p_0^{s+t}) = O(R^{4n-2(sk+jt)}/p_0).$$

Now (3.10) follows. From this and (viii) we get for $j = 0, \ldots, n-1$

$$(v^{2n+\alpha} {}_R{}^{4n}/p_0)^{(j)} = \sum_{\ell=0}^{j} \binom{j}{\ell} (v^{2n+\alpha})^{(\ell)} ({}_R{}^{4n}/p_0)^{(j-\ell)}$$

$$= 0(\sum_{\ell=1}^{j} v^{2n-\ell+\alpha} {}_R{}^{8n-2j}/p_0^2) + 0(v^{2n+\alpha} {}_R{}^{4n-2j}/p_0)$$

$$= 0(v^{2n-j+\alpha} {}_R{}^{4n-2j}/p_0) .$$

In I_2, putting $p_{n-j} = r_{n-j} + q_{n-j}$, and using (3.11), (iii) and c)

$$| \int (v^{2n+\alpha} {}_R{}^{4n}/p_0)^{(j-s)} r_{n-j} f^{(s)} f^{(j)}|$$

$$\leq K \int v^{2n+\alpha-j+s} {}_R{}^{2j+2s} |f^{(s)} f^{(j)}|$$

(3.12)

$$\leq K \int v^{j+s+\alpha} {}_R{}^{2j+2s} |f^{(s)} f^{(j)}|$$

$$\leq K(F_j F_s)^{\frac{1}{2}} .$$

From i), in I_2,

$$| \int (v^{2n+\alpha} {}_R{}^{4n}/p_0)^{(j-s)} q_{n-j} f^{(s)} f^{(j)}|$$

$$= | \int h_{n-j} \{(v^{2n+\alpha} {}_R{}^{4n}/p_0)^{(j-s)} f^{(s)} f^{(j)}\}^{(n-j)}|$$

(3.13)

$$= |\sum_{k=0}^{n-j} \sum_{t=0}^{k} \int h_{n-j} (v^{2n+\alpha} {}_R{}^{4n}/p_0)^{(n-s-k)} f^{(s+k-t)} f^{(j+t)}|$$

$$\leq K \sum_{k=0}^{n-j} \sum_{t=0}^{k} \int v^{s+k+j+\alpha} {}_R{}^{2(s+k+j)} |f^{(s+k-t)} f^{(j+t)}|$$

$$\leq K \sum_{k=0}^{n-j} \sum_{t=0}^{k} (F_{s+k-t} F_{j+t})^{\frac{1}{2}}$$

on using iv), (3.11) and c). Hence, from (3.12) and (3.13)

$$|I_2| \leq K \sum_{\substack{\ell,m=0 \\ \ell+m<2n}}^{n} (F_\ell F_m)^{\frac{1}{2}}$$

(3.14)

$$\leq \varepsilon_1 F_n + K(\varepsilon_1) \sum_{\ell=0}^{n-1} F_\ell$$

for any $\varepsilon_1 > 0$. As in (3.12) we get

(3.15)

$$|I_3| \leq K \sum_{m=1}^{n-1} (F_m F_n)^{\frac{1}{2}}$$

$$\leq \varepsilon_2 F_n + K(\varepsilon_2) \sum_{m=1}^{n-1} F_m .$$

In I_4, we have from ii)

$$- \int (v^{2n+\alpha} {}_R{}^{4n}/p_0) q_n f^2$$

$$\leq - \int (v^{2n+\alpha} {}_R{}^{4n}/p_0) h_n^{(n)} f^2$$

$$= (-1)^{n+1} \sum_{j=0}^{n} \binom{n}{j} \int h_n (v^{2n+\alpha} R^{4n}/p_0)^{(j)} (f^2)^{(n-j)}$$

$$= (-1)^{n+1} \int h_n (v^{2n+\alpha} R^{4n}/p_0)^{(n)} f^2$$

$$+ (-1)^{n+1} \sum_{j=0}^{n-1} \sum_{k=0}^{n-j} \binom{n}{j} \binom{n-j}{k} \int h_n (v^{2n+\alpha} R^{4n}/p_0)^{(j)}$$

$$f^{(k)} f^{(n-j-k)}$$

$$= (-1)^{n+1} \int h_n v^{2n+\alpha} (R^{4n}/p_0)^{(n)} f^2$$

$$+ (-1)^{n+1} \sum_{s=0}^{n-1} \binom{n}{s} \int h_n (v^{2n+\alpha})^{(n-s)} (R^{4n}/p_0)^{(s)} f^2$$

$$+ (-1)^{n+1} \sum_{j=0}^{n-1} \sum_{k=0}^{n-j} \binom{n}{j} \binom{n-j}{k} \int h_n (v^{2n+\alpha} R^{4n}/p_0)^{(j)}$$

$$f^{(k)} f^{(n-j-k)}$$

$$= J_1 + J_2 + J_3$$

say. From v) and iv), for arbitrary ε_3 and $\varepsilon_4 > 0$,

$$J_1 \leq \varepsilon_3 \int v^{2n+\alpha} h_n^2 [(R^{4n}/p_0)^{(n)}]^2 f^2 + (K/\varepsilon_2) v^{2n+\alpha} f^2$$

(3.16)
$$\leq \frac{4\varepsilon_3}{(1+\delta)} \int v^{2n+\alpha} R^{4n} p_0^{-2} h_n^2 \{K + R^{4n} p_0^{-1} r_n\} f^2 + (K/\varepsilon_3) \int v^{\alpha} f^2$$

$$\leq \varepsilon_4 \int (R^{4n} r_n/p_0) v^{2n+\alpha} f^2 + (K/\varepsilon_4) F_0.$$

From (3.10), viii), iv) and a)

(3.17)
$$J_2 \leq K F_0$$

and from (3.11) and c), for any $\varepsilon_5 > 0$

$$J_3 \leq K \sum_{j=0}^{n-1} \sum_{k=0}^{n-j} (F_k F_{n-j-k})^{\frac{1}{2}}$$

(3.18)
$$\leq \varepsilon_5 F_n + K(\varepsilon_5) \sum_{m=1}^{n-1} F_m.$$

Hence

$$I_4 \leq -(1-\varepsilon_4) \int (R^{4n} r_n/p_0) v^{2n+\alpha} f^2$$

(3.19)
$$+ \varepsilon_5 F_n + K(\varepsilon_4, \varepsilon_5) \sum_{m=0}^{n-1} F_m.$$

Since $\sum_{\ell=1}^{n} (-1)^{\ell} \binom{n}{\ell} = -1$, we get

$$I_5 = -\int (v^{2n+\alpha}\, R^{4n}/p_0)^{(n)}{}_{p_0} f f^{(n)}$$

$$= -\int v^{2n+\alpha}\, (R^{4n}/p_0)^{(n)}{}_{p_0} f f^{(n)}$$

$$- \sum_{j=0}^{n-1} \binom{n}{j} \int (v^{2n+\alpha})^{(n-j)}\, (R^{4n}/p_0)^{(j)}{}_{p_0} f f^{(n)}$$

$$\le -\int v^{2n+\alpha}\, (R^{4n}/p_0)^{(n)}{}_{p_0} f f^{(n)}$$

(3.20)
$$+ K(F_0 F_n)^{\frac{1}{2}}$$

$$\le \int \{[p_0 (R^{4n}/p_0)^{(n)}]^2\, v^{2n+\alpha} f^2\}^{\frac{1}{2}}\, \{v^{2n+\alpha}(f^{(n)})^2\}^{\frac{1}{2}} + \varepsilon_6 F_n + K/\varepsilon_6 F_0$$

$$\le 2(1+\delta)^{-\frac{1}{2}} \int \{(R^{4n}/p_0 r_n + K)v^{2n+\alpha} f^2\}^{\frac{1}{2}}\, \{v^{2n+\alpha}\, R^{4n}(f^{(n)})^2\}^{\frac{1}{2}}$$

$$+ \varepsilon_6 F_n + (K/\varepsilon_6)F_0$$

$$\le 1/\varepsilon_7 (1+\delta)^{\frac{1}{2}} \int (R^{4n}/p_0 r_n)v^{2n+\alpha} f^2 + (\varepsilon_6 + \varepsilon_7/(1+\delta)^{\frac{1}{2}})F_n$$

$$+ K(\varepsilon_6, \varepsilon_7)F_0$$

on using v) and for arbitrary $\varepsilon_6, \varepsilon_7 > 0$.

We therefore get from (3.9), (3.14), (3.15), (3.19) and (3.20)

$$F_n \le K \int v^{2n+\alpha}|fMf|$$

(3.21)
$$+ (\varepsilon_1 + \varepsilon_2 + \varepsilon_5 + \varepsilon_6 + \varepsilon_7/(1+\delta)^{\frac{1}{2}})F_n$$

$$-(1 - \varepsilon_4 - \frac{1}{\varepsilon_7 (1+\delta)^{\frac{1}{2}}})\int (R^{4n}/p_0\, r_n)\, v^{2n+\alpha} f^2$$

$$+ K(\varepsilon_1, \varepsilon_2, \varepsilon_4, \varepsilon_5, \varepsilon_6, \varepsilon_7) \sum_{m=0}^{n-1} F_m \ .$$

If we now choose $1/(1+\delta)^{\frac{1}{2}} < \varepsilon_7 < 1$ and other ε_i sufficiently small we get, since $-(R^{4n}/p_0)r_n \le K$ from (v)

$$F_n \le K \int v^{2n+\alpha}|fMf| + K \sum_{m=0}^{n-1} F_m$$

and hence (3.5). The proof of (3.8) is similar.

3.3 Lemma. Under the hypothesis of Lemma 3.2, given any $\varepsilon > 0$ there exists a number $K(\varepsilon) > 0$, independent of I, such that

(3.22)
$$F_{n-m} \le \varepsilon F_n + K(\varepsilon)F_0, \quad m = 1,2,\ldots,n$$

(3.23)
$$G_{n-m} \le \varepsilon G_n + K(\varepsilon)G_0, \quad m = 1,2,\ldots,n \ .$$

(3.24) $\qquad F_j \le \epsilon G_0 + K(\epsilon)F_0, \quad j = 0,1,\dots,n$

(3.25) $\qquad G_j \le \epsilon \int v^{8n+\alpha}(M^2 f)^2 + K(\epsilon)G_0, \quad j = 0,1,\dots,n.$

Proof. We will prove the first inequality only as the proof of the second one is similar. The last two follow from (3.6) and (3.8). First we establish by induction that

(3.26) $\qquad F_j \le \epsilon F_{j+1} + K(\epsilon)F_0, \quad j = 1,2,\dots,n-1.$

This holds for $j = 1$ by (3.5). Suppose it holds for $j = 1,2,\dots,s$. Then, from (3.5)

$$F_{s+1} \le \epsilon_1 F_{s+2} + K(\epsilon_1)F_s \le \epsilon_1 F_{s+2} + K(\epsilon_1)\{\epsilon_2 F_{s+1} + K(\epsilon_2)F_0\}.$$

Choosing ϵ_1, ϵ_2 appropriately we get

$$F_{s+1} \le \epsilon F_{s+2} + K(\epsilon)F_0$$

and (3.24) follows.

We now prove (3.22) again using induction. It is true for $m = 1$ by (3.26). From (3.26) and the induction hypothesis

$$F_{n-m-1} \le \epsilon_1 F_{n-m} + K(\epsilon_1)F_0 \le \epsilon_1\{\epsilon_2 F_n + K(\epsilon_2)F_0\} + K(\epsilon_1)F_0$$

which yields (3.22).

4. Notes and comments. Theorem 2.1 is due to Evans-Zettl [37]. We are indebted to W. D. Evans for permission to include this as yet unpublished result in these notes. It includes many of the known higher order limit-point criteria for real expressions. Some of the authors who have dealt with this topic are Atkinson [2], Brown-Evans [8], Devinatz [10-17], Eastham [23,24], Evans [32,33], Everitt [40], Hinton [71,73], Kauffman [81,82].

Among the authors who have obtained sufficient conditions on the coefficients for $d(M) = k$ with $k > n$ and M a real symmetric expression of order $2n$ for $n > 1$ are: Eastham [22,27], Devinatz [16,18], Devinatz-Kaplan [20], Fedorjuk [59,60], Hinton [70,72], Kauffman [82], Kogan-Rofe-Beketov [86], Naimark [97], Orlov [100], Walker [124], Wood [134].

The complex coefficient case has been studied by Hinton [72], Kogan and Rofe-Beketov [86], Kumar [89], McLeod [96].

The positive definite case is when all the coefficients p_k of M in (1.1) satisfy $p_k \geq 0$. Until quite recently all known examples in the positive definite case were limit-point. This gave rise to a conjecture that all positive definite expressions are in the limit-point case. This was settled in the negative by Kauffman [82] for $n > 2$. The case $n = 2$ is still open.

The case $n = 1$ of Example (2.1) has been studied in Atkinson-Eastham-McLeod [3].

CHAPTER V

THE DEFICIENCY INDEX PROBLEM FOR POLYNOMIALS IN SYMMETRIC DIFFERENTIAL EXPRESSIONS.

1. Introduction.

In this chapter we investigate the relationship between the deficiency indices of M and M^k, $k = 1,2,3,\ldots$ where M is a given symmetric expression of any order and M^k is the kth power of M - powers of symmetric expressions are symmetric.

For example if M is the second order expression

$$My = -(py')' + qy$$

one can ask: How is $d(M^2)$ or $d(M^3)$ related to $d(M)$? Is it possible to deduce $d(M^2)$ from $d(M)$ or vice versa?

A complete description of the possible sequences $d(M), d(M^2), d(M^3),\ldots$ for any symmetric M with real coefficients will be given in this and the next chapter. In section 4 we deal with the general limit-circle case i.e. when $d(M)$ is maximal. In section 5 we show that $d(M^k) \geq kd(M)$ and in 6 a necessary and sufficient condition for equality is developed. Explicit conditions on the coefficients for the determination of $d(M^k)$ are given in chapter 7.

Below we wish to consider products of differential expressions M and N. To make this notion precise let

$$My = p_m y^{(m)} + p_{m-1} y^{(m-1)} + \ldots + p_0 y$$
$$Ny = q_n y^{(n)} + q_{n-1} y^{(n-1)} + \ldots + q_0 y.$$

Certainly My is defined for any function y in C^m. So $(MN)y = M(Ny)$ is defined for Ny in C^m and Ny is in C^m if q_j and $y^{(n)}$ are in C^m. Assuming q_j in C^m, $j = 0,\ldots,n$ we define MN by

$$(MN)y = M(Ny) \quad \text{for all } y \text{ in } C^{m+n}.$$

In particular M^2 will be defined - as an operator on C^m - if the coefficients p_j are in C^m, $j = 0,\ldots,m$. For M^3 to be defined in the above (classical) manner we can assume that $p_j \in C^{2m}$ for $j = 0,\ldots,m$ and so on.

Clearly if the coefficients are all C^∞ functions we can consider any power of M or, more generally any polynomial in M.

For coefficients which are sufficiently differentiable (e.g. C^∞) so that $(MN)^+$ can be formed we have $(MN)^+ = N^+M^+$ - see [21, pp. 1289]. In particular this shows that powers of M are symmetric if M is. Since also $(M + N)^+ = M^+ + N^+$ and $(\lambda N)^+ = \bar{\lambda} N$ for any complex number λ, it is clear that $p(M)$ is symmetric for any symmetric M and any real polynomial p.

Since we are particularly interested in powers of symmetric differential expressions we list a few examples for later reference.

(1.1)   Let $My = (py')' + qy$. Then

$$M^2 y = (p^2 y'')'' + ((2pq + pp'')y')' + (q^2 + pq'' + p'q')y$$

$$M^3 y = (p^3 y''' + ((4p^2 p'' + 3p^2 q)y'')''$$
$$+ ((p^2 p^{(4)} + 5pp'q' + 3qpp'' + 2pp''p'' + 4p^2 q'' + 3pq^2)y')'$$
$$+ \{[p(q^2 + pq'' + p'q')']' + q(q^2 + pq'' + p'q')\}y.$$

In the important special case when $p(t) = 1$ for $t \geq 0$ we have:

$$M^2 y = y^{(4)} + (2qy')' + (q^2 + q'')y$$

2. Minimal and maximal operators of product expressions.

For the remainder of this chapter, whenever we consider products of differential expressions, we will assume that the coefficients are sufficiently differentiable so that these products can be formed in the manner described above. For simplicity it may be assumed that the coefficients are C^∞ functions.

For given differential expressions M_1, M_2, ..., M_n how are the minimal operators (see chapter 2 for the definitions) $T_0(M_1 M_2 \cdots M_k)$ and $T_0(M_1)T_0(M_2)\cdots T_0(M_k)$ related? Here we investigate this relationship which plays an important role in our subsequent development.

The main "tool" in this study is a result of Glassman. This was used by him in his celebrated paper [62] to construct examples of real symmetric expressions M of order $2n$ such that $d(M) = k$ for any k, $n \leq k \leq 2n$, $n = 2,3,4\ldots$.

Although this result is discussed in chapter 2 we repeat it here for the convenience of the reader.

2.1 Theorem. (Glassman [62]). Let A_j be a closed operator with dense domain in a Hilbert space H and suppose that it has closed range and finite deficiency,

$j = 1, \ldots, k$. Then $\Pi_{j=1}^{k} A_j$ is a closed operator with dense domain, closed range and

$$\text{def. } \Pi_{j=1}^{k} A_j = \sum_{j=1}^{k} \text{def. } A_j.$$

Proof. See Theorem A in [62].

As an application of this theorem we get

2.2 Theorem. Suppose M_j is a regular differential expression – not necessarily symmetric – on the interval $[0, \infty)$ such that its minimal operator $T_0(M_j)$ has closed range, $j = 1, \ldots, k$.   Then the product operator $\Pi_{j=1}^{k} T_0(M_j)$ is closed, has dense domain, closed range and

(i) $$\text{def. } \Pi_{j=1}^{k} T_0(M_j) = \sum_{j=1}^{k} \text{def. } T_0(M_j);$$

and

(ii) $$T_0(M_1 M_2 \cdots M_k) \subseteq \Pi_{j=1}^{k} T_0(M_j).$$

Remark. Note that the containment in part (ii) may be proper i.e. the operators $T_0(M_1 M_2 \cdots M_k)$ and $T_0(M_1) T_0(M_2) \cdots T_0(M_k)$ are not equal in general. This was established by Glassman [62]. Chaudhuri and Everitt in [9] give an example of an expression M such that $T_0(M^2) \neq (T_0(M))^2$.

Proof of Theorem 2.1. Let $L = T_0(M_1 M_2 \cdots M_k)$ and $R = T_0(M_1) T_0(M_2) \cdots T_0(M_k) = \Pi_{j=1}^{k} T_0(M_j)$.

Since each minimal operator $T_0(M_j)$ has no eigenvalues and has closed range by assumption we may conclude that $\lambda = 0$ is a regular type point of $T_0(M_j)$. Also, by Theorem 2.3.13, $T_0(M_j)$ has dense domain and finite deficiency. Therefore, by Theorem 2.1, R is closed, has $\lambda = 0$ as a regular type point, has dense domain and

$$\text{def. } R = \sum_{j=1}^{k} \text{def. } T_0(M_j);$$

this establishes part (i).

For part (ii) note that L and R agree on the C_0^∞ functions, i.e. $Ly = Ry = M_1 M_2 \cdots M_k y$ for all y in C_0^∞. By definition L is the minimal closed extension obtained from $M_1 M_2 \cdots M_k$ restricted to C_0^∞. Since R is closed it follows that $L \subseteq R$. This completes the proof of part (ii).

Some consequences of Theorem 2.2 are now mentioned as Corollaries.

2.3 Corollary. Let M_j be a regular, not necessarily symmetric, differential

expression on $[0, \infty)$ for $j = 1,\ldots,k$. If all solutions of the differential

equations $M_j y = 0$ and $M_j^+ y = 0$ are in $L^2(0, \infty)$ for $j = 1,\ldots,k$; then all

solutions of $(M_1 M_2 \ldots M_k) y = 0$ and of $(M_1 M_2 \ldots M_k)^+ y = 0$ are in $L^2(0, \infty)$.

Proof. Let n_j = order M_j = order M_j^+, $j = 1,\ldots,k$. Then def. $T_0(M_j^+) = n_j =$

def. $T_0(M_j)$. Since all of the solutions of $M_j y = 0$ and $M_j^+ y = 0$ are in

$L^2(0, \infty)$, the minimal operators $T_0(M_j)$ and $T_0(M_j^+)$ have closed range. This

can be seen from the fact that the inverse operators $[T_0(M_j)]^{-1}$ and $[T_0(M_j^+)]^{-1}$

are compact (on their domains of definition) and hence continuous since they are

generated by integral operators with Hilbert-Schmidt kernels. These statements

can be established by adapting the standard proofs in [21, XIII 3 and 4] or

[97, pp. 195-200] for the fact that resolvents of self-adjoint extensions of the

minimal operator $T_0(M)$ - for M a symmetric expression in the maximal deficiency

case - are compact operators.

By Theorem 2.1 we have, using the notation from the proof of Theorem 2.2,

$$\text{def. } L \geq \text{def. } R = \sum_{j=1}^{k} n_j = \text{order } (M_1 \ldots M_k) = \text{order}(M_1 \ldots M_k)^+.$$

Thus def. L = order of $(M_1 \ldots M_k)^+$ and consequently all solutions of $(M_k^+ \ldots M_1^+)^+ y$

$= (M_1 \ldots M_k) y = 0$ are in $L^2(0, \infty)$. Repeating this argument with M_j^+ replaced

by M_j we conclude that all solutions of $(M_1 \ldots M_k)^+ y = 0$ are also in $L^2(0, \infty)$.

A more direct proof of Corollary 2.3, one which does not appeal to Glassman's

abstract result, can be obtained by using the variation of parameters formula.

In the special case when $M_j = M$ for all $j = 1,\ldots,k$ and M is symmetric

the converse of Corollary 2.3 holds: If all solutions of $M^k y = 0$ are in $L^2(0, \infty)$

then all solutions of $My = 0$ are in $L^2(0, \infty)$ since these are also solutions of

$M^k y = 0$. In general if all solutions of $(M_1 \ldots M_k) y = 0$ are in $L^2(0, \infty)$ then all

solutions of $M_k y = 0$ are in $L^2(0, \infty)$. Also if all solutions of the adjoint

equation $(M_1 \ldots M_k)^+ y = 0$ are in $L^2(0, \infty)$ then it follows immediately that all

solutions of $M_1^+ y = 0$ are in $L^2(0, \infty)$. So in particular for $k = 2$ we have

established the simple

2.4 Corollary. Suppose M_1, M_2 and $M_1 M_2$ are all regular symmetric expressions.

Then the product M_1M_2 is in the limit-circle case if and only if both M_1 and M_2 are in the limit-circle case.

As a special example of Corollary 2.4 when both M_1 and M_2 are of the second order and in the limit-circle case, then the fourth order expression M_1M_2 has deficiency index 4.

In connection with the application of Theorem 2.2 to get information about the deficiency indices of <u>symmetric</u> differential expressions we note that the product of symmetric expressions is not symmetric in general. However any power of a symmetric expression is symmetric and so called symmetric products such as

$$M_1M_2M_1, \ M_1M_2M_3M_2M_1, \ \text{etc.}$$

of symmetric expressions are symmetric.

<u>2.5 Corollary</u>. Suppose M_j is a regular symmetric differential expression on $[0, \infty)$ such that its minimal operator $T_0(M_j)$ has closed range, $j = 1,\ldots,k$. Let $M = M_1 \ldots M_{k-1} \ M_k \ M_{k-1} \ldots M_1$. Then M is a regular symmetric expression, each of the expressions M, M_j has equal deficiency indices which we denote by $d(M), d(M_j)$, respectively, $j = 1,\ldots,n$ and

$$d(M) \geq d(M_k) + 2 \sum_{j=1}^{k-1} d(M_j).$$

Proof. By part (ii) of Theorem 2.2 we have

$$T_0(M) \subseteq T_0(M_1) \ \cdots \ T_0(M_{k-1})T_0(M_k)T_0(M_{k-1}) \ \cdots \ T_0(M_1).$$

From this containment, the definition of the deficiency index of an operator and part (i) of Theorem 2.2 follows that

$$\text{def. } T_0(M) \geq \text{def. } T_0(M_k) + 2 \sum_{j=1}^{k-1} \text{def. } T_0(M_j).$$

It remains only to note (see chapter 2) that def. $T_0(M) = d_+(M) = d_-(M) = d(M)$ to obtain the conclusion of Corollary 2.5.

Remark. Assume the hypothesis of Corollary 2.5 hold. Given the deficiency indices $d(M_j)$ of M_j some restrictions are placed on $d(M)$. If all the $d(M_j)$'s are minimal then these restrictions are vacuous in the sense that no information not already available from the general classification results is obtained. At the other extreme when all the $d(M_j)$'s are maximal, then $d(M)$ is determined - it must also be maximal. Between these two extremes $d(M)$ is not determined but many

possibilities allowed by the general classification results are ruled out. To

illustrate these points we mention two simple examples.

   1. Let $k = 2$ and assume each M_j is of order 2. If $d(M_1) = 2 = d(M_2)$,

then $d(M_1 M_2 M_1) = 6$; this is the limit-circle case for M_1, M_2 and $M_1 M_2 M_1$. If

$d(M_1) = 2$ and $d(M_2) = 1$, then $d(M_1 M_2 M_1)$ is either 5 or 6. If $d(M_1) = 1$ and

$d(M_2) = 2$, then $d(M_1 M_2 M_1)$ is either 4 or 5, since $d(M_1 M_2 M_1) = 6$ is not possible

as this would imply $d(M_1) = 2$.

   2. Let $k = 2$ and M_1 be of order 2, M_2 of order 4. Then $M = M_1 M_2 M_1$ has

order 8 and so the possibilities, according to the classification results for equal

deficiency indices, for $d(M)$ are 4, 5, 6, 7 and 8. Now $d(M_1) = 2$ and $d(M_2) = $

rules out 4, 5 and 6 for $d(M)$. If $d(M_1) = 2$ and $d(M_2) = 2$ or $d(M_1) = 1$ and

$d(M_2) = 4$, then only 4 and 5 are ruled out by Corollary 2.5. (In the latter case

8 is also ruled out by Corollary 2.4.)

   Corollary 2.5 can be used "in reverse": Knowledge of $d(M)$ for

$M = M_1 \ldots M_{k-1} M_k M_{k-1} \ldots M_1$ can be used to gain information about the $d(M_j)$'s. If,

with the hypothesis of Corollary 2.5, M is limit-point then each M_j must be

limit-point. Also if, for example, the order of M_1 is 2 and M_2 is 4 and

$d(M_1 M_2 M_1) = 5$, then $d(M_1) = 1$ and $d(M_2) = 3$ or 2.

   When is the containment in part (ii) of Theorem 2.2 actually equality?

2.6 Theorem. Let M_1, M_2, $\ldots$, M_k be regular, not necessarily symmetric,

differential expressions on $[0, \infty)$. Suppose $T_0(M_j)$ has closed range for

$j = 1, \ldots, k$. Then

(2.1) $\qquad T_0(M_1 \ldots M_k) = \Pi_{j=1}^{k} T_0(M_j)$

if and only if the following partial separation condition is satisfied:

(2.2) $\qquad$ f and $(M_k^+ M_{k-1}^+ \ldots M_1^+) f$ in $L^2(0, \infty)$ together imply that

$(\Pi_{j=1}^{r} M_j^+) f$ is in $L^2(0, \infty)$ for $r = 1, 2, \ldots, k-1$.

Furthermore $T_0(M_1 \ldots M_k) = \Pi_{j=1}^{k} T_0(M_j)$ if and only if def. $T_0(M_1 \ldots M_k) = $
$\sum_{j=1}^{k}$ def. $T_0(M_j)$.

We shall say that the product $M_1 \ldots M_k$ is partially separated whenever (2.2) holds.

Consider $T_0(M_j)$ - the minimal operator corresponding to the expression M_j, $j = 1, \ldots, k$. Is the minimal operator $T_0(M_1 \ldots M_k)$ of the product expression $M_1 \ldots M_k$ the product of the minimal operators $T_0(M_1) \ldots T_0(M_k)$? In general the answer is no. Basically the reason for this is that the domain of $T_0(M_1 \ldots M_k)$ may be smaller than the domain of the product operator $T_0(M_1) \ldots T_0(M_k)$. The partial separation condition is precisely the condition needed to make these domains equal. The proof given below is a technical verification of this statement. To make it work in general we need the closed range hypothesis on $T_0(M_j)$. As we will see in section 4 below this hypothesis can be eliminated if $M_j = M$ for $j = 1, \ldots, k$ and M is symmetric.

Proof. Let $L = T_0(M_1 \ldots M_k)$ and $R = T_0(M_1) \ldots T_0(M_k)$. Taking adjoints in part (ii) of Theorem 2.2 and using the fact that $A \subseteq B$ implies $B^* \subseteq A^*$ we obtain $R^* \subseteq L^*$. Using $(AB)^* \supseteq B^* A^*$ and the relationships between minimal and maximal operators discussed in chapter 2 we have

$$
\begin{aligned}
T_1(M_k{}^+) \, T_1(M_{k-1}{}^+) \ldots T_1(M_1{}^+) &= (T_0(M_k))^* (T_0(M_{k-1}))^* \ldots (T_0(M_1))^* \\
&\subseteq (T_0(M_1) \ldots T_0(M_{k-1}) T_0(M_k))^* \\
&\subseteq T_0(M_1 \ldots M_{k-1} M_k)^* \\
&= T_1((M_1 \ldots M_{k-1} M_k)^+) = T_1(M_k{}^+ M_{k-1}{}^+ \ldots M_1{}^+).
\end{aligned}
$$

Thus (2.1) will follow from $R \subseteq L$. To show this we establish that $L^* \subseteq R^*$. This suffices since $A^{**} = A$ for a closed operator A. Since L^* and R^* agree on the C_0^∞ functions, it suffices to prove that $D(L^*) \subseteq D(R^*)$. Let $f \in D(L^*)$. Then f is in $L^2(0, \infty)$ and $(M_k{}^+ \ldots M_1{}^+)f$ exists a.e. and is in $L^2(0, \infty)$. From the partial separation hypothesis it follows that

$$
M_1{}^+ f, \ (M_2{}^+ M_1{}^+)f, \ \ldots, \ (M_{k-1}{}^+ \ldots M_1{}^+)f
$$

are all in $L^2(0, \infty)$. Hence $f \in D(T_1(M_1{}^+))$ and $T_1(M_1{}^+)f = M_1{}^+ f \in D(T_1(M_2{}^+))$ and so on to give, finally, $f \in D(R^*)$. Hence $L^* \subseteq R^*$ and $L = R$.

On the other hand, suppose $L = R$. Then

$$T_0(M_1 \ldots M_k) * = T_1(M_k^+ \ldots M_1^+) = R* =$$

$$[T_0(M_1) \ldots T_0(M_k)]* \supseteq [T_0(M_k)]* \ldots [T_0(M_1)]* =$$

$$T_1(M_k^+) \ldots T_1(M_1^+).$$

Next we show that $T_1(M_k^+ \ldots M_1^+)$ and $T_1(M_k^+) \ldots T_1(M_1^+)$ have the same index. (The index $k(A)$ of an operator A is its nullity minus its deficiency.) By [63, theorem IV.2.7, p. 103], $k(T_1(M_k^+) \ldots T_1(M_1^+) = k(T_1(M_k^+)) + \ldots + k(T_1(M_1^+))$. Using the fact that $k(A*) = -k(A)$ - see [63] - we get $k(T_1(M_k^+ \ldots M_1^+)) = -k(T_0(M_1 \ldots M_k))$. From the hypothesis $L = R$ and using theorem IV.2.7 from [63] again we have:

$$-k(T_0(M_1 \ldots M_k)) = -k(T_0(M_1) \ldots T_0(M_k))$$

$$= - k(T_0(M_1)) + \ldots + k(T_0(M_k))$$

$$= k(T_1(M_k^+) + \ldots + k(T_1(M_1^+)).$$

Hence (see [76], p. 233, problem 5.8)

$$T_1(M_k^+ \ldots M_1^+) = T_1(M_k^+) \ldots T_1(M_1^+).$$

From the equality of the domains of these two operators the partial separation condition (2.2) follows readily.

The furthermore remark in Theorem 2.6 follows from the fact that if A and B are Fredholm operators, i.e. closed operators with closed range and finite nullity and deficiency and $A \subseteq B$, then index A = index B implies $A = B$ - see [76, p. 233, problem 5.8].

The hypotheses that $T_0(M)$ have closed range plays an important role in Theorem 2.2 and Theorem 2.6. In general it is difficult to determine what expressions M have this property. Below we list some who do.

Let N be a regular differential expression on $[0, \infty)$. Then the minimal operator $T_0(N)$ has closed range if any one of the following conditions is satisfied:

1. All solutions of the differential equations $Ny = 0$ and $N^+ y = 0$ are in $L^2(0, \infty)$; in the symmetric case $N = N^+$ this is equivalent to N being in the

limit-circle case.

2. $N = M - \lambda$ where M is a symmetric differential expression and λ is a non-real complex number. This will allow us, in section 5 below, to remove the assumption that the $T_0(M_j)$ have closed range in both Theorems 2.2 and 2.6 for the special case when $M_j = M$ for all j and M is symmetric.

3. $Ny = \sum_{j=0}^{n} (-1)^j (p_j y^{(j)})^{(j)}$ with $p_n > 0$, $p_j(t) \geq 0$, $j = 1, 2, \ldots, n-1$ and $p_0(t) \geq \delta > 0$ for $t \geq 0$. Integration by parts shows that, with $T_0 = T_0(N)$,

$$(T_0 y, y) = \int_0^\infty \bar{y} \, T_0 y = \sum_{j=0}^{n} \int_0^\infty p_j |y^{(j)}|^2 \geq \delta \int_0^\infty |y|^2$$

for all y in C_0^∞ and hence all $y \in D(T_0(N))$. Combining this with the Schwartz inequality yields

$$||T_0 y|| \geq \delta ||y|| \qquad \text{(for all } y \in D(T_0)\text{)}.$$

From this follows that $\lambda = 0$ is a regular type point of the operator T_0 and that T_0 has closed range.

In connection with 3 above it is interesting to note that until recently no expressions of this type were known for which the minimal operator was not in the limit-point case. At the Dundee conference on ordinary and partial differential equations in March 1976 R. M. Kauffman announced a class of examples of 6th order expressions of type 3 with deficiency index 4 and another class of 6th order such expressions with deficiency index 5. Kauffman also obtained such examples for order greater than 6 and gave a complete classification of the positive definite case with polynomial coefficients.

As a further illustration of Corollary 2.5 let M_1 and M_2 be of type 3 above both of order 6. If $d(M_1) = 5 = d(M_2)$, then $d(M) \geq 15$ for $M = M_1 M_2 M_1$. Since $d(M)$ cannot be 18 (then $d(M_1) = 6$) it must be 15, 16 or 17. Thus the possibilities for $d(M)$ have been reduced in number from 10 to 3. Furthermore, by Theorem 2.6, $d(M) = 15$ if and only if $M_1 M_2 M_1$ is partially separated.

3. The relationship between $d(M^k)$ and $d(M)$.

In this section we consider the special case of Theorems 5.2.2 and 5.2.6 when $M_j = M$ for all j and M is symmetric. For this case we will be able to eliminate the troublesome assumption that the minimal operator have closed range.

Consequently our results here are quite general.

<u>3.1 Theorem</u>. Let M be a regular symmetric expression on $[0, \infty)$. Then

$$d_+(M^k), \quad d_-(M^k) \geq r[d_+(M) + d_-(M)]$$

for $k = 2r$ and

$$d_+(M^k) \geq (r + 1)d_+(M) + rd_-(M),$$
$$d_-(M^k) \geq rd_+(M) + (r + 1)d_-(M)$$

for $k = 2r + 1$, $r > 0$.

Proof. Let $M_j = M - w_j$, $j = 1, \ldots, k$. If the w_j are the nth roots of the complex number $i = \sqrt{-1}$, then we can write

$$T_0(M^k) - iI = T_0(\Pi_{j=1}^k (M - w_j)).$$

By Theorem 2.2 and example 2 following Theorem 2.6

$$\begin{aligned} \text{def. } T_0(M^k - i) &= \text{def. } T_0((M - w_1)(M - w_2) \ldots (M - w_k) \\ &\geq \text{def. } \Pi_{j=1}^k T_0(M - w_j) \\ &= \Sigma_{j=1}^k \text{def. } T_0(M - w_j). \end{aligned}$$

From this and the observations that (1) for $k = 2r$ there are exactly r of the w_j's in the upper and exactly r in the lower, half-planes of the complex plane and (2) for $k = 2r + 1$ there are exactly $r + 1$ of the w_j's in the upper and r in the lower half-planes; it follows that for $k = 2r$

$$d_-(M^k) \geq r[d_+(M) + d_-(M)]$$

and for $k = 2r + 1$

$$d_-(M) \geq rd_+(M) + (r + 1)d_-(M).$$

The other two inequalities are obtained by replacing i by $-i$ and proceeding similarly.

Strict inequality can occur in Theorem 3.1. Examples to illustrate this will be given in section 5.

As an immediate consequence of Theorem 3.1 and the general classification results for deficiency indices we obtain two Corollaries.

3.2 Corollary. If, for some $k > 1$, $d_+(M^k)$ or $d_-(M^k)$ take on the minimum value possible according to the general classification result, then both $d_+(M)$ and $d_-(M)$ are minimal, i.e. M is in the limit-point case. In particular, if some power of M is limit-point, then M must be limit-point.

We have already remarked above that the converse of the last statement is false in general. It turns out that M can be limit-point with M^2 not limit-point etc. In section 5 we will give examples to illustrate this and all of the other possibilities.

Theorem 3.1 and Corollary 3.2 can be strengthened by replacing M^k by $p(M)$ where p is any polynomial of degree k with real coefficients. (In fact the coefficients need not be real, but then we're out of the symmetric case.) In chapter VI it will be shown that $d_+(p(M)) = d_+(M^k)$ and $d_-(p(M)) = d_-(M^k)$ for any k-th degree polynomial p with real coefficients and positive leading coefficient.

Proof of Corollary 3.2. We consider four cases.

(i) Suppose k is even, $k = 2m$ and n, the order of M, is odd, $n = 2r + 1$, $r > 0$. If $d_+(M^k) = mn$, then $mr + m(r + 1) \le md_+(M) + md_-(M) \le mn$. Hence $d_+(M) = r$ and $d_-(M) = r + 1$. The argument for the case $d_-(M^k) = mn$ is entirely similar.

The other cases follow similarly.

Corollary 3.3. If M is limit-circle, then M^k is limit-circle for any $k \ge 2$.
Proof. This is immediate from Theorem 3.2 and the general classification results.

In the next section we will see that the converse of Corollary 3.3 is true and more generally that M is limit-circle if and only $p(M)$ is limit-circle for any polynomial p with real coefficients.

Since powers of limit-point expressions are not, in general, limit-point we now take up the question of when they are. More generally one can ask: When does $d(M^k) = kd(M)$ for, say, real symmetric expressions M?

Although the concept of partial separation for a power M^k is a special case of this concept for products discussed in section 2 we repeat it here for convenience.

3.4 Definition. Let M be a symmetric differential expression and let k be a positive integer greater than 1. We say that M^k is partially separated if f and $M^k f$ in $L^2(0, \infty)$ together imply that $M^r f$ is in $L^2(0, \infty)$ for $r = 1,2,\ldots,k-1$.

The reader is reminded that, by our convention, the assumption $M^k f$ in $L^2(0, \infty)$ includes the hypothesis that $f^{(s-1)}$ is locally absolutely continuous where s is the order of M^k. This means that $M^k f$ exists a.e. on $(0, \infty)$ and is locally integrable.

3.5 Theorem. Let M be a regular symmetric expression and k a positive integer greater than 1. Then

(i) If M^k is partially separated, then

$$d_+(M^k) = d_-(M^k) = r[d_+(M) + d_-(M)]$$

when $k = 2r$ is even, and

$$d_+(M^k) = (r + 1)d_+(M) + rd_-(M), \quad N_-(M^k) = rd_+(M) + (r + 1)d_-(M)$$

when $k = 2r + 1$, $r > 0$ is odd.

Conversely,

(ii) If, for $k = 2r$ even, either

$$d_+(M^k) = r[d_+(M) + d_-(M)] \quad \text{or}$$
$$d_-(M^k) = r[d_+(M) + d_-(M)]$$

and, for $k = 2r + 1$, $r > 0$, odd, either

$$d_+(M^k) = (r + 1)d_+(M) + rd_-(M) \quad \text{or}$$
$$d_-(M^k) = rd_+(M) + (r + 1)d_-(M);$$

then M^k is partially separated.

An important special case is mentioned as

3.6 Corollary. If $d_+(M) = d_-(M) = d(M)$, as is always the case when M is real, then $d(M^k) = kd(M)$ if and only if M^k is partially separated.

It is this characterization which we will use in chapter 7 to get sufficient conditions on the coefficients for all powers M^k (and hence all real polynomials as we will see in chapter 6) to be limit-point.

Proof of Theorem 3.5. Let $M_j = M - w_j$, $j = 1, \ldots, k$ where $w_1, \ldots, w_k$ are the kth roots of the complex number $i = \sqrt{-1}$. Then each minimal operator $T_0(M - w_j)$ has closed range - see example 2 following Theorem 2.6. Also we can write

$$M^k - i = \Pi_{j=1}^k (M - w_j).$$

Hence by Theorem 2.6 we have

$$T_0(M^k - i) = T_0(\Pi_{j=1}^k (M - w_j))$$
$$= \Pi_{j=1}^k T_0(M - w_j))$$

if and only if M^k is partially separated.

Now assume that M^k is partially separated. By Theorem 2.1 and the above

$$\text{def. } T_0(M^k - i) = \text{def. } \Pi_{j=1}^k T_0(M - w_j)$$
$$= \Sigma_{j=1}^k \text{def. } T_0(M - w_j).$$

For k even, say $k = 2r$, exactly r of the w_j's are in the upper half-plane and exactly r in the lower and for $k = 2r + 1$, $r > 0$ we have exactly $r + 1$ of the w_j's in the upper and r in the lower half-planes. Therefore we can conclude that

(a) for $k = 2r$: $d_+(M^k) = r[d_+(M) + d_-(M)]$ and

(b) for $k = 2r + 1$: $d_+(M^k) = (r + 1)d_+(M) + rd_-(M)$.

Replacing i by $-i$ and proceeding as above we get the corresponding results for $d_-(M^k)$.

For part (ii) note that either (a) or (b) implies that def. $T_0(M^k - i) =$ def. $[(T_0(M))^k - iI]$. From this and Theorem 2.2 it follows that

$$T_0(\Pi_{j=1}^k (M - w_j) = T_0(M^k - i) = \Pi_{j=1}^k T_0(M - w_j)$$

and hence from Theorem 2.6 that M^k is partially separated. The rest follows similarly.

4. The limit-circle case for powers.

For expressions of order greater than two there are very few sufficient conditions known for the limit-circle case. There is even some scarcity of examples. The next theorem provides a simple method for generating class of limit-circle expressions of all orders.

4.1 Theorem. Let M be a regular symmetric expression. Then M is limit-circle if and only if $p(M)$ is limit-circle for any polynomial p with real coefficients.

Although a part of this theorem was proven in section 2 we give an independent and elementary proof here. First two simple lemmas are established.

4.2 Lemma. Let M be a differential expression, λ a complex number and p a real polynomial. If $My = \lambda y$, then $p(M)y = p(\lambda)y$.

Proof. Let $p(x) = a_k x^k + a_{k-1} x^{k-1} + \ldots + a_1 x + a_0$, a_j real.

Then

$$
\begin{aligned}
p(M)y &= (a_k M^k + a_{k-1} M^{k-1} + \ldots + a_1 M + a_0)y \\
&= a_k M^k y + a_{k-1} M^{k-1} y + \ldots + a_1 My + a_0 y \\
&= a_k \lambda^k y + a_{k-1} \lambda^{k-1} y + \ldots + a_1 \lambda y + a_0 y = p(\lambda)y.
\end{aligned}
$$

4.3 Lemma. If $\lambda_1, \lambda_2, \ldots, \lambda_k$ are distinct complex numbers and, for each $j = 1,2,\ldots,k$, z_m^j are linearly independent solutions of $My = \lambda_j y$, then the set of functions $\{z_q^j, \ j = 1,\ldots,k \ \text{and} \ q = 1,\ldots,m\}$ is linearly independent.

Proof. Suppose there exist complex numbers c_q^j, $j = 1,\ldots,k$ $q = 1,\ldots,m$ such that $y_1 + y_2 + \ldots + y_k = 0$ where $y_j = c_1^j z_1^j + \ldots + c_m^j z_m^j$ for $j = 1,\ldots,k$. Then $M(y_1 + y_2 + \ldots + y_k) = \lambda_1 y_1 + \lambda_2 y_2 + \ldots + \lambda_m y_m = 0$. Repeated applications of M yield:

$$
\lambda_1^r y_1 + \lambda_2^r y_2 + \ldots + \lambda_k^r y_k = 0, \quad r = 0,1,\ldots,k-1.
$$

Thus we obtain a $k \times k$ linear homogeneous system of equations whose coefficients form a Van der Monde matrix. Since Van der Monde determinants are non-zero we have

$$
y_i = 0 \quad \text{for} \quad i = 1,\ldots,k.
$$

Therefore, by the linear independence of

$$
z_1^j, \ z_2^j, \ \ldots, \ z_m^j \quad \text{we get} \quad c_q^j = 0 \quad \text{for} \quad j = 1,\ldots,k, \quad q = 1,\ldots,m.
$$

Proof of Theorem 4.1. Choose a complex number C such that the roots of the real polynomial equation

$$
p(x) = C,
$$

say $\lambda_1, \lambda_2, \ldots, \lambda_k$ are distinct. For each $j = 1, \ldots, k$ let $z_1^j, z_2^j, \ldots, z_m^j$ be a fundamental set of (linearly independent) solutions of

(4.1) $My = \lambda y$

with $\lambda = \lambda_j$. By Lemma 4.2 each z_q^j is a solution of

(4.2) $\qquad p(M)y = Cy$

and by Lemma 4.3 the z_q^j, $j = 1,\ldots,k$ $q = 1,\ldots,m$ form a fundamental set of solutions of (4.2). Hence $p(M)$ is limit-circle since each z_q^j is in $L^2(0, \infty)$.

On the other hand if $p(M)$ is limit-circle, we choose C and the λ_q's as above and conclude that all solutions of (4.1) with $\lambda = \lambda_1$ are in $L^2(0, \infty)$. Hence M is limit-circle.

5. Examples.

Here we shall determine which sequences $\{r_m\}_{m=1}^{\infty}$ of positive integers occur as the deficiency indices associated with the sequence $\{M^m\}_{m=1}^{\infty}$ of powers of some $2n^{th}$ order formally symmetric differential expression M which has real-valued infinitely differentiable coefficients on the interval $[0, \infty)$ and is regular at 0. [See chapter 2 for the definition of the deficiency index.]

We know from chapter 2 that for any expression M as described above, the deficiency index r of M satisfies the inequality $n \leq r \leq 2n$. As a special case of our construction we shall obtain for each integer r in this interval a formally symmetric expression of the form $e^x M_1$, where M_1 has constant coefficients which has deficiency index r. These expressions provide a more elementary alternative to the class of examples with this property constructed by Glazman [62, p. 350].

Several general restrictions on the sequence $\{r_m\}$ of deficiency indices associated with the sequence $\{M^m\}$ are known. One follows immediately from the application of the inequality of the previous paragraph to M^m: for each integer m,

(a) $\qquad mn \leq r_m \leq 2mn$.

In chapter 6 we will establish that if $r_0 = 0$ and $s_m = r_m - r_{m-1}$, then

(b) $\qquad$ the sequence $\{s_m\}_{m=1}^{\infty}$ is non-decreasing.

Equivalently, the graph of the piecewise linear function with vertices at $(0, 0)$ and the lattice points (m, r_m) is convex.

It follows easily from (a) and (b) that $s_1 \geq n$ and that the sequence $\{s_m\}$

is eventually constant with a value $s_\infty \leq 2n$. In particular, if $s_1 = r_1 = 2n$,
then $s_m = 2n$ for each m. If $n = 1$ it follows from (a) and (b) that if
$r_k > k$ for any k, then the entire sequence $\{r_m\}$ is completely determined
($s_m = 1$ for $m < 2k - r_k$, and $s_m = 2$ for $m \geq 2k - r_k$).

For $n > 1$ it is in general very far from being the case that a single r_k
together with (a) and (b) suffice to determine other r_m either for $m < k$ or
$m > k$. One might hope, therefore, to supplement (a) and (b) with additional
general information which would at least further restrict the possible sequences
$\{r_m\}$. We shall show, however, that no such additional rules exist, that is, every
sequence $\{r_m\}$ which satisfies (a) and (b) is the sequence of deficiency indices
associated with the powers of some $2n^{\text{th}}$ order M.

We shall do this by constructing a class of differential expressions with
the property that every such sequence of positive integers occurs as the sequence
of deficiency indices associated with the powers of a member of the class.

First, however, we shall discuss the particularly simple expressions in this
class which provide an alternative to the examples of Glassman. The analysis for
these expressions is quite similar to that for the general case but without the
technical complications which will arise there.

It will be convenient to write out expressions as products of first-order
expressions. We shall write D for the differentiation operator d/dx and M^+
for the formal (Lagrange) adjoint of M. Now let M be the $2n^{\text{th}}$ order formally
symmetric expression

(5.1) $$M = M_1^+ M_2^+ \cdots M_n^+ e^x M_n \cdots M_1,$$

where for each $j = 1, 2, \ldots, n$,

$$M_j = D + a_j = D + i_j + 1/(j+2)$$

and each i_j is either 0 or 1. Note that $M = e^x M_1$ for some expression M_1 with
constant coefficients. Set z equal to the number of values of j for which
$i_j = 0$. Then we shall show that the deficiency index r of M is equal to $n + z$.
Thus any r with $n \leq r \leq 2n$ can be obtained by choosing the i_j's so that
$z = r - n$.

To establish this assertion we remark first that for each of the first-order expressions M_j in (5.1) and for each complex-valued infinitely differentiable function f with compact support in $(0, \infty)$,

$$\left| \int_0^\infty \overline{f}\, M_j(f) \right| \geq a_j \int_0^\infty |f|^2.$$

Hence by Schwarz's inequality .

(5.2) $$\int_0^\infty |M_j(f)|^2 \geq a_j^2 \int_0^\infty |f|^2$$

for all such f. A similar inequality, with a_j replaced by 1 holds for the expression $E(f) = e^x f$. It follows that any composition of the M_j's and E also satisfies an inequality of the form (5.2) with a_j replaced by some positive c. In particular, M has this property. Thus the minimal operator corresponding to M has closed range in the space $L^2(0, \infty)$ of square-integrable functions. This implies that the deficiency index of M is equal to the number of linearly independent solutions of $M(y) = 0$ which lie in $\mathbf{L}^2(0, \infty)$. (See chapter 2.)

We shall now construct a basis $B = \{y_1, \ldots, y_{2n}\}$ of the space of solutions of $L(y) = 0$ by taking y_1 to be a non-trivial solution of $M_1(y) = 0$, y_2 to be a solution of $M_2 M_1(y) = 0$ which is not a solution of $M_1(y) = 0$ and so forth, working from right to left in (5.1).

We begin by setting $y_1(x) = e^{-a_1 x}$. The remainder of the construction of B is based on the elementary remark that if a, b and k are real numbers with $a + b \neq 0$ and $k \neq 0$, then

(5.3) $$y' + ay = ke^{bx} \text{ has a solution of the form } y = ce^{bx}.$$

To apply (5.3), note that if y_2 is a solution of $M_1(y) = y' + a_1 y = e^{-a_2 x}$ then $M_1(y_2) \neq 0$ and $M_2 M_1(y_2) = M_2(e^{-a_2 x}) = 0$. By (5.3) we may take $y_2(x) = ce^{-a_2 x}$. (Of course we may now set $c = 1$ without affecting the fact that $M_2 M_1(y_2) = 0$ and $M_1(y_2) \neq 0$.) Similarly, to find y_3 it suffices to solve $M_2 M_1(y) = e^{-a_3 x}$. Set $u = M_1(y)$. It follows first from applying (5.3) to $M_2(u) = e^{-a_3 x}$ that we may take $u(x) = ke^{-a_3 x}$, and then from applying (5.3) to $M_1(y) = u = ke^{-a_3 x}$ that we may take $y_3(x) = e^{-a_3 x}$. In the same way we find from $j - 1$ applications of (5.3) that we may take $y_j(x) = e^{-a_j x}$ for each $j \leq n$.

The situation is slightly different for $j > n$. We wish y_{n+1} to satisfy

$$e^x M_n \ldots M_1(y) = e^{a_n x}$$

or $M_n \ldots M_1(y) = e^{(a_n - 1)x}$. Now a repetition of the previous argument leads to

$y_{n+1}(x) = e^{(a_n - 1)x}$. Similarly, for each $k = 1, 2, \ldots, n$ we find from repeated

application of (5.3) and one multiplication by e^{-x} at the appropriate time that

$$y_{2n-k+1}(x) = e^{(a_k - 1)x}.$$

Thus we have

(5.4) $\qquad B = \{e^{-a_1 x}, \ldots, e^{-a_n x}, e^{(a_1 - 1)x}, \ldots, e^{(a_n - 1)x}\}.$

By the choice of the a_j's, the $2n$ exponents are all distinct and different
from 0. It follows that a non-trivial linear combination of elements of B is in
$L^2(0, \infty)$ if and only if each element of B which appears in the linear combination
is itself in $L^2(0, \infty)$. The number of elements of B which are in $L^2(0, \infty)$ is
equal to n (for the first n elements of B) plus the number of a_j's such
that $a_j - 1 < 0$. But this second number is just the number z of a_j's for
which i_j is equal to 0. Thus the deficiency index of (5.1) is $n + z$ and the
example is complete.

The justification of (5.4) could be shortened somewhat by observing that M_1,
$M_2, \ldots, M_n$ commute. However, we have emphasized the repeated application of
(5.3) instead because this method is essentially the same as the argument to be
given in the proof of Theorem 5.1 where the M_j's do not commute.

The deficiency indices of the powers of (5.1) could be computed in a similar
fashion by extending B to a basis B_m of the space of solutions of $M^m(y) = 0$
which still consists of exponential functions. If the class of expressions is
enlarged by allowing the integer parts i_j of the a_j's to be any non-negative
integer, then an analysis similar to the one above would establish that
$s_m = r_m - r_{m-1}$ is equal to n plus the number of a_j's with integer parts at
most $m - 1$. In particular, $s_m = 2n$ for all sufficiently large m. Thus any
sequence $\{r_m\}$ of positive integers satisfying (a) and (b) and the additional
condition that $s_\infty = 2n$ occurs as the sequence of deficiency indices associated

with the powers of a differential expression of this simple type.

In order to obtain sequences of deficiency indices for which $s_\infty < 2n$ it is necessary to modify the form of the expressions discussed above by including some factors M_j of the form $D + a(x + 1)$. The observation (5.3) must then be replaced by a lemma (to be stated precisely below) concerning first-order differential equations with polynomial coefficients. With these modifications we shall now obtain our main result.

<u>5.1 Theorem</u>. Every sequence $\{r_m\}_{m=1}^\infty$ of positive integers which satisfies (a) and (b) is a sequence of deficiency indices associated with the powers $\{M^m\}_{m=1}^\infty$ of some $2n^{th}$ order formally symmetric differential expression M.

Proof. Let $\{r_m\}_{m=1}^\infty$ be a sequence of positive integers which satisfies (a) and (b). The differential expression M such that r_m is the deficiency index of M^m for each m will have the form

$$(5.5) \qquad M = M_1^+ \ldots M_n^+ e^x M_n \ldots M_1 ,$$

where each M_j is a first-order expression of the form $M_j = D + Q_j$ and the Q_j's are polynomials of degree at most one which we shall now construct from the sequence $\{r_m\}$.

Set $N = s_\infty - n$. (Recall that $s_\infty = r_m - r_{m-1}$ for all large m.) Then $0 \leq N \leq n$. Define a finite non-decreasing sequence $\{i_j\}_{j=1}^N$ by setting $i_j = m - 1$ for

$$s_{m-1} - n < j \leq s_m - n.$$

(We take $s_0 = n$.) Thus the number of i_j's which are less than m is exactly $s_m - n$ for each positive integer m.

Now define $Q_1, \ldots, Q_n$ as follows.

$$Q_j(x) = i_j + 1/(j + 2), \qquad j = 1, \ldots, N$$
$$Q_j(x) = j(x + 1), \qquad\qquad j = N + 1, \ldots, n.$$

Note that $Q_j(x) < m - 1/2$ for all x if $j = 1, 2, \ldots, s_m - n$ and that $Q_j(x)$ is eventually greater than m for all other values of j. It is also easy to see that neither

$$Q_j + Q_k, \quad 1 \leq j, \quad k \leq n, \quad \text{nor} \quad Q_j - Q_k, \quad j \neq k,$$

can be eventually equal to an integer.

We assert first that for each m the minimal operator of M^m has closed range or, equivalently, that there are positive constants c_m such that for each complex-valued infinitely differentiable function f, $||M^m(f)|| \geq c_m ||f||$. (The symbol $||\cdot||$ here refers to the usual norm in $L^2(0, \infty)$.) As in the discussion of the example in (5.1), each M^m is a composition of M_j's and the expression $E(f) = e^x f$ so that it suffices to establish such an inequality for each individual M_j. This was done in (5.1) for $M_1, \ldots, M_N$. Moreover, the same calculation shows that (5.2) also holds for $M_{N+1}, \ldots, M_n$ with a_j replaced by j. Thus the minimal operator of each M^m does have closed range so that for each m the deficiency index of M^m is equal to the number of linearly independent solutions of $M^m(y) = 0$ in $L^2(0, \infty)$.

We consider first the situation for m = 1. Define polynomials $P_1, \ldots, P_{2n}$ by

$$P_j(x) = - \int_0^x Q_j,$$
$$P_{n+j}(x) = (\int_0^x Q_j) - x, \qquad j = 1, 2, \ldots, n.$$

We assert that there is a basis B of the solution space of $M(y) = 0$ of the form

(5.6) $$B = \{R_1 e^{P_1}, \ldots, R_{2n} e^{P_{2n}}\},$$

where each R_j is a differentiable function which is asymptotic to a power of x.

Assume for the moment that this is true. By the choice of the Q_j's the exponents are all distinct and eventually bounded away from each other and 0 by a non-zero multiple of x. Thus the R_j's cannot affect whether an element of B is in $L^2(0, \infty)$. Each of the first n elements of B is in $L^2(0, \infty)$. The first $s_1 - n$ elements of the rest of B are also in $L^2(0, \infty)$ since the integer part i_j of Q_j is equal to 0 if and only if $j \leq s_1 - n$. No non-trivial linear combination of the remaining elements of B is in $L^2(0, \infty)$. Thus, if (5.6) is a basis for the solution space of $M(y) = 0$, then there are exactly $n + (s_1 - n) = s_1$ linearly independent solutions of $M(y) = 0$ in $L^2(0, \infty)$ and it follows that the deficiency index of M is equal to $s_1 = r_1$ as asserted.

We shall now argue as above to establish (5.6). In place of (5.3) we shall need the following lemma.

5.2 Lemma. Let P and Q be polynomials with $P' + Q$ not identically 0, and let R be a differentiable zero-free function on $[0, \infty)$ such that $R'(x)/R(x) \to 0$ as $x \to \infty$. Then the equation

$$(5.7) \qquad (D + Q)y = Re^P$$

has a solution $y = R_0 e^P$ on $[0, \infty)$ where R_0 is differentiable and zero-free, and $R'_0(x)/R_0(x) \to 0$ as $x \to \infty$. Moreover, R_0 is asymptotic to a power of x if R is.

Proof. For $x \geq 0$, define

$$F(x) = P(x) + \int_0^x Q.$$

Then F is a non-constant polynomial. Suppose first that $F' = P' + Q$ is eventually positive. Define

$$G(x) = \int_0^x Re^F.$$

Then $y(x) = G(x) \exp(- \int_0^x Q)$ is a solution of (5.7) on $[0, \infty)$ and $y = R_0 e^P$ with $R_0 = Ge^{-F}$. The first two assertions concerning R_0 clearly hold. To establish the others we shall first show that if H is any function, defined and differentiable for all large x, such that $(H'/G')(x) \to 0$ as $x \to \infty$, then $(H/G)(x) \to 0$ as $x \to \infty$.

To see this, note that $G''/G' = R'/R + F' > k > 0$ for all sufficiently large x. For such values of x, $|G'(x)|e^{-kx}$ is strictly increasing, so that both $|G'(x)|$ and $|G(x)|$ tend to infinity with x. If H is as above, choose a so that $|H'/G'|(x) < \varepsilon/2$ for $x \geq a$ and then $b \geq a$ so that $|H(a)/G(x)| < \varepsilon/2$ for $x \geq b$. Then for $x \geq b$,

$$|H/G|(x) \leq |H(a)/G(x)| + |(\int_a^x H')/G(x)| < \varepsilon.$$

As a first application, set $H = (G'/F') - G$. $H(x)$ exists and is differentiable for all x greater than the largest zero of F'. Then, recalling that $G''/G' = R'/R + F'$,

$$H'/G' = (R'/RF) + 1 - (F''/F') - 1.$$

Since F is a polynomial and $R'/R \to 0$, this implies that H'/G', and hence also H/G, approach 0. Equivalently,

(5.8)
$$\lim_{x \to 0}(GF'/G')(x) = 1.$$

Since $GF'/G' = R_0 F'/R$, this establishes that R_0 is asymptotic to a power of x if R is.

Now set $H = G' - GF'$. Then

$$H'/G' = R'/R - GF''/G'$$

$$= R'/R - (GF'/G')(F''/F')$$

so that again $H'/G' \to 0$ as $x \to \infty$. Thus

(5.9)
$$\lim_{x \to \infty}(R'_0/R_0)(x) = \lim_{x \to \infty}(H/G)(x) = 0$$

and the proof of the lemma is complete in this case.

If F' is eventually negative, then for all large x, $(Re^F)'/Re^F < -k < 0$ so that $|Re^F|(x)e^{kx}$ is decreasing. Define $G(x) = -\int_x^\infty Re^F$ and set

$$y(x) = G(x) \exp\left(-\int_0^x Q\right)$$

and $R_0 = Ge^{-F}$ as before. We must modify the previous argument slightly, since the justification for (5.8) and (5.9) no longer applies. However, $G''/G' < -k < 0$ eventually so that both G and G' approach 0 exponentially as $x \to \infty$. Hence both GF' and $G' - GF'$ approach 0 as $x \to \infty$. Thus (5.8) and (5.9) now follow from an application of l'Hospital's rule and the proof of the lemma is complete.

We now return to the proof of the theorem. We have seen that to complete the proof for $m = 1$ it suffices to establish that the space of solutions of $M(y) = 0$ has a basis of the form (5.2). This may now be done in exactly the same way as above, except that each use of (5.3) is now replaced by an appeal to the lemma. Each such appeal is justified because in each case $P' + Q$ is equal to either $Q_j - Q_k$ or $Q_j + Q_k - 1$ for some j and k and none of these polynomials is identically 0. This completes the proof for $m = 1$.

For $m \geq 2$ we may proceed similarly to extend the basis B successively to a nested sequence $\{B_m\}$ of bases and of the spaces of solutions of $M^m(y) = 0$.

The 2n additional elements in B_2 and not in B arise as solutions of

$$M_1M(y) = 0, \quad M_2M_1M(y) = 0, \quad \ldots, \quad M^2(y) = 0.$$

The only difference between the calculation of a solution of $M_j \ldots M_1(y) = 0$ and a solution of $M_j \ldots M_1M(y) = 0$ (aside from the fact that 2n additional applica tions of the lemma are required) is that in the second case one additional factor of e^x appears on the left side and is moved to the right as in the calculation of y_{n+1} above in connection with (5.1). This has the effect that the exponents of the new elements in B_2 are of the form

$$(5.10) \qquad P_1(x) - x, \quad \ldots, \quad P_{2n}(x) - x,$$

that is, x has been subtracted from each of the exponents in (5.2). Again the choice of $Q_1, \ldots, Q_n$ ensures that the 4n exponents of the elements of B_2 are distinct and non-zero so that the number of linearly independent solutions of $M^2(y) = 0$ which are in $L^2(0, \infty)$ is equal to the number of elements of B_2 with eventually negative exponents. In (5.10) the first n terms are negative, as are those of the last n terms for which $P_{n+j}(x) < x$ or equivalently $Q_j(x) < 2$. Now $Q_j(x) < 2$ if $j \leq s_2 - n$. Otherwise $Q_j(x)$ is eventually greater than 2. Hence, since s_1 of the elements of B are in $L^2(0, \infty)$, the deficiency index of M is

$$s_1 + n + (s_2 - n) = s_1 + s_2 = r_2.$$

Similarly, it is the case for each m that the number of $L^2(0, \infty)$ solutions which are in B_m and not in B_{m-1} is equal to n plus the number of values of j for which $Q_j(x) < m$, that is, to $n + (s_m - n) = s_m$. It follows that the deficiency index of M^m is equal to $s_1 + s_2 + \ldots + s_m = r_m$ and the proof is complete.

It is clear from the proof that the deficiency indices of the powers of (5.1) can be calculated in this way for many other choices of $Q_1, \ldots, Q_n$. The only properties of the Q_j's actually used were (1) that for any integer i each function $Q_j + Q_k - i$ and $Q_j - Q_k - i$ $(j \neq k)$ is eventually bounded away from 0 and (2) that each Q_j is sufficiently smooth for the proof of the lemma to be valid.

6. Notes and comments.

The results in section 3 are due to Zettl [137, 138]. Some special cases had been discovered previously by Everitt and Giertz [47-53]. The first paper to study the relationship between $d(M)$ and $d(M^k)$ seems to be the 1969 paper [9] of Chaudhuri and Everitt where $d(M^2)$ was investigated for $My = -(py')' + qy$. In particular they gave an example of such an M with $d(M) = 1$ and $d(M^2) = 3$. Also the partial separation condition has its roots in this paper, although the terminology was introduced later by Everitt and Giertz. The extensions of the results on M^k in section 3 to products $M_1 M_2 \ldots M_k$ given in section 2 is due to Everitt and Zettl [58]. The examples of positive definite type expressions $(p_j \geq 0)$ not in the limit-point case mentioned at the end of section 2 were developed by Kauffman [82]. Still open is the question of the existence of such expressions in the limit-circle case.

The discussion of the limit-circle case in section 4 is elementary.

The examples in section 5 are due to Read [110]. It is interesting to note that for

$$My = \sum_{j=0}^{n} (p_j y^{(j)})^{(j)}$$

all possible cases for $d(M)$ i.e. $n \leq d(M) \leq 2n$ can be realized with $p_j(t) = c_j e^t$, $j = 0, \ldots, n$ just by choosing the constants c_j appropriately.

CHAPTER VI

APPLICATIONS OF PERTURBATION THEORY

1. Introduction.

In this chapter we discuss the theory of the mean deficiency index for differential operators, not necessarily formally symmetric, which are polynomials in lower order differential operators.

The first idea that one might have is that, if M is limit-point, then all polynomials in M are also limit-point. We have seen that this is false; L^2 need not even be limit-point. However, there is a very definite relationship between the mean deficiency index of M^j and that of M^k, when $M = M^+$. Furthermore, any two polynomials of the same degree in M have the same mean deficiency index. Finally, the rules relating the deficiency index of polynomials of different degree are the best possible--examples were given in 5.5 to show that all possibilities not specifically excluded by these rules actually do occur.

It will be convenient to begin with the results on polynomials. We permit polynomials with complex coefficients, both for completeness and because the proof of the rules for powers of a symmetric expression   uses results about complex polynomials in M. Thus, even to deal with symmetric expressions, we must study non-symmetric expressions.

The reader who wishes to know the range of possibilities for the relationship of the mean deficiency index of M^j with that of M^k should read the rule of section 3.

2. Polynomials in a symmetric expression.

Let M be a regular ordinary differential expression with infinitely differentiable coefficients on $[a, \infty)$. We will discuss polynomials in M by showing that the term of highest degree is dominant. What is meant by "dominant" is related in the following lemma.

Lemma 2.1. Let H be a symmetric operator in a Hilbert space h. Suppose that f_n is a sequence in the domain of H^{2j}, with $||f_n|| = 1$ and $|(H^i f_n, f_n)|$

approaching infinity for some $i < 2j$. Then $(H^i f_n, f_n)/(H^{2j} f_n, f_n)$ converges to zero as n approaches infinity.

Proof. H has a self-adjoint extension (perhaps to a larger Hilbert space.) Let P_λ be a spectral decomposition of this self-adjoint extension.

Then, by the spectral theorem,

$$(H^i f_n, f_n) = \int_{-\infty}^{\infty} \lambda^i d(p_\lambda f_n, f_n) = \int_{-\infty}^{\infty} \lambda^i du_n,$$

where μ_n is a positive regular Borel measure on $(-\infty, \infty)$, with $\mu_n(-\infty, \infty) = 1$.

Now, if $1/p + 1/q = 1$, and f is in $Lp(\mu_n)$, and g is in $Lq(\mu_n)$, $|\int fg du_n| \leq ||f||_p ||g||_q$ by Hölder's inequality. But $||f||_p = |\int |f|^p du_n|^{1/p}$. Let f be $|\lambda|^i$, and g be 1, with $p = 2j/i$. Then $\int |\lambda|^i du_n \leq (\int |\lambda|^{2j} du_n)^{i/2j}$. Thus, if $\int \lambda^i du_n$ approaches infinity with n, it follows that $\int |\lambda|^i du_n$ approaches infinity, so $\int \lambda^{2j} du_n$ approaches infinity. The conclusion follows.

Corollary 2.2. Let H be a symmetric operator in a Hilbert space h. Suppose that f_n is a sequence with $||f_n|| = 1$ in domain H^{2j}, such that, for some $i < 2j$, $|(H^i f_n, f_n)|$ approaches infinity. Then, for all $k < 2j$, $(H^k f_n, f_n)/(H^{2j} f_n, f_n)$ approaches zero as n approaches infinity.

Lemma 2.3. Suppose that M is a regular, formally symmetric ordinary differential expression on $[a, \infty)$ with C^∞ coefficients. Let $N = M^j + \sum_0^{j-1} g_i; M^i$, with each g_i a bounded, C^∞, complex valued function. Then there is a $K > 0$ such that, for all $i \leq j$, $||M^i f|| \leq K(||f|| + ||Nf||)$ for all f in domain $T_0(N)$.

Proof. We first show that there is a $K > 0$ such that $K(||f|| + ||Nf||) \geq ||M^i f||$ for all f in $C_0^\infty(a, \infty)$, where $1 \leq i \leq j$.

Suppose that the above statement were not true. Then there is a sequence f_n in C_0^∞ with $||f_n|| = 1$ and $||M^i f_n|| > n(||f_n|| + ||Nf_n||)$. The operator $T_0(M)$, being symmetric, has a self-adjoint extension H, perhaps to a larger Hilbert space. We then have

(2.1) $$||H^i f_n||^2 > n^2(||f_n||^2 + 2||f_n|| \, ||Lf_n|| + ||Lf_n||^2).$$

Since $||f_n|| = 1$, it follows that $(H^{2i} f_n, f_n) > n$. Thus $(H^k f_n, f_n)/(H^{2j} f_n, f_n)$ converges to zero as n approaches infinity, for any $1 \leq k \leq 2j-1$. Therefore

$||Nf_n||^2/(H^{2j}f_n, f_n)$ approaches 1. Dividing both sides of (2.1) by $(H^{2j}f_n, f_n)$, we obtain a contradiction. Therefore we know that there is a $K > 0$ such that, for all f in $C_0^\infty(a, \infty)$, $K(||f|| + ||Nf||) \geq ||M^i f||$. We must extend this inequality to all f in domain $T_0(N)$.

Recall that $T_0(N)$ is the closure of its restriction to $C_0^\infty(a, \infty)$. Thus, if $<f, Nf>$ is in the graph of $T_0(N)$, there is a sequence $<f_n, Nf_n>$ in $L^2[a, \infty) \times L^2[a, \infty)$, with f_n in $C_0^\infty(a, \infty)$, and with $<f_n, Nf_n>$ converging to $<f, Nf>$ in the product space. We just finished proving an inequality that guarantees that $M^i f_n$ is a Cauchy sequence in L^2 for any $1 \leq i \leq j$. Since $T_0(M^i)$ is a closed operator, $T_0(M^i)f_n$ converges to $T_0(M^i)f$ for each i. Therefore $K(||f|| + ||Nf||) \geq ||M^i f||$ for all f in domain $T_0(N)$. The lemma is proved.

Lemma 2.4. Suppose that M is a regular formally symmetric ordinary differential expression with C^∞ coefficients on $[a, \infty)$. Let j be any fixed positive integer. Then, for any $B > 0$, there is an $\alpha > 0$ such that, for any $\lambda > \alpha$ and any j-tuple of bounded C^∞ functions with

$$\sum_0^{j-1}||g_m||_\infty \leq B, \quad T_0(M^j + \sum_0^{j-1} g_k M^k + \lambda z)$$

has closed range, where $z = \pm(-1)^{1/2}$.

Proof. If this lemma were not true, there would be a sequence f_n in $C_0^\infty(a, \infty)$ with $||f_n|| = 1$ and $||(N_n + \alpha_n z)f_n||$ converging to zero, where

$$N_n = M^j + \sum_{k=0}^{j-1} g_{n,k} M^k, \quad \sum_{k=0}^{j-1}||g_{n,k}||_\infty \leq B, \quad \text{and} \quad \alpha_n > n.$$

To see this, observe that $T_0(Q)$ is always a closed operator, and that range $T_0(Q)$ is closed if and only if $(T_0(Q))^{-1}$ is a bounded linear transformation from range $T_0(Q)$ into $L^2[a, \infty)$, where Q is any regular ordinary differential expression with differentiable coefficients on $[a, \infty)$. Since $T_0(Q)$ is the closure of its restriction to $C_0^\infty(a, \infty)$, the f_n may be chosen in $C_0^\infty(a, \infty)$.

Let $||(N_n + \alpha_n z)f_n|| = c_n$. We know by hypothesis that c_n converges to zero. Since $||\alpha_n z f_n|| \geq n$, it follows that, for some $k \leq j$, $||M^k f_{n_i}||$ must converge to infinity for some subsequence f_{n_i}. Without loss of generality we may assume that f_{n_i} is the original sequence f_n.

By Corollary 2.2, we now see that, for any $k < j$, $(M^{2k}f_n, f_n)/(M^{2j}f_n, f_n)$ converges to zero. Thus $||M^k f_n||/||M^j f_n||$ converges to zero for any $k < j$. It then follows that, since $c_n/||M^j f_n||$ converges to zero, and $||N_n f_n||/||M^j f_n||$ converges to one, then $\alpha_n/||M^j f_n||$ also converges to one. Thus, for any $k < j$, $||M^k f_n||/\alpha_n$ converges to zero.

However, $(M^j f_n, f_n)$ is real, for each n. It follows from the Schwarz inequality that $|(g_{n,k}M^k f_n, f_n)|/\alpha_n$ converges to zero for $k < j$. Therefore the imaginary part of $(N_n f_n, f_n)/\alpha_n$ converges to zero. It follows that the imaginary part of $((N_n + \alpha_n z)f_n, f_n)/\alpha_n$ converges to one. This is impossible, since $||(N_n + \alpha_n z)f_n||$ converges to zero and $||f_n|| = 1$. The lemma is proved.

We need an elementary lemma next.

<u>Lemma 2.5.</u> Suppose that M_1 and M_2 are regular ordinary differential expressions with differentiable coefficients on $[a, \infty)$, and that, for some $K > 0$, and all f in $C_0^\infty(a, \infty)$ $||M_2 f|| \le K(||f|| + ||M_1 f||)$. Then domain $T_0(M_2)$ contains domain $T_0(M_1)$.

Proof. $T_0(M_1)$ is the closure of its restriction to $C_0(a, \infty)$. However, if f_n and $M_1 f_n$ are Cauchy, it is easy to see that $M_2 f_n$ is Cauchy. Furthermore, since $T_0(M_2)$ is closed, $M_2 f_n$ converges to $M_2 f$ where f is the limit of the sequence f_n. The conclusion follows.

<u>Lemma 2.6.</u> Let M be a regular, formally symmetric ordinary differential expression with C^∞ coefficients on $[a, \infty)$. Let $N = M^j + \sum_0^{j-1} g_j M^j$, where the g_j are bounded C^∞ complex valued functions. Then domain $T_0(N) = $ domain $T_0(M^j)$, and, for some $K > 0$, $||M^i f|| \le K(||f|| + ||Mf||)$ for all f in domain $T_0(N)$ and all $i \le j$.

Proof. If f is in domain $T_0(M^j)$, then there is a sequence f_n in $C_0^\infty(a, \infty)$ with f_n converging to f and $M^j f_n$ converging to $M^j f$. Since $T_0(M)$ has a self-adjoint extension (perhaps to a larger Hilbert space), it then follows from Lemma 2.1 that $M^i f_n$ is a Cauchy sequence for $i < j$, which must then converge to $M^i f$ since $T_0(M^i)$ is a closed operator. Thus f is in the domain of $T_0(N)$.

Now, if f is in the domain of $T_0(N)$ and f is not in domain $T_0(M^i)$, for some $i \leq j$, then by the preceding lemma there is a sequence f_n in $C_0^\infty(a, \infty)$ with $||f_n|| = 1$ and $||M^i f_n|| > n(||f_n|| + ||Nf_n||)$. It then follows from Corollary 2.2 that, for large n, $||M^j f_n|| \geq n(||f_n|| + ||Nf_n||)$, so that, by Corollary 2.2, $||Nf_n||/||M^j f_n||$ converges to one. This is a contradiction, and we have proved that f is in domain $T_0(M^j)$, and also that, for some $K > 0$ and all f in $C_0^\infty(a, \infty)$, $||M^i f|| \leq K(||f|| + ||Nf||)$. But since this inequality holds for all f in C_0^∞, we may pass to the closure to see that it also holds on domain $T_0(N)$.

Thoerem 2.7. Let M be a regular, formally symmetric ordinary differential expression with C^∞ coefficients on $[a, \infty)$. Let $N = M^j + \sum_0^{j-1} g_i M^i$, where the g_i are bounded C^∞ functions. Then there is an $\alpha > 0$ such that, for all $\lambda > \alpha$, the nullity of $T_1(M^j + \lambda z) = $ nullity $T_1(N^+ + \lambda z)$, where $z = \pm(-1)^{1/2}$.

Proof. Select α as in Lemma 2.4, so that for all $\lambda > \alpha$ and all bounded C^∞ functions $\{h_i\}_{i=0}^{j-1}$ such that $\sum_0^{j-1} ||h_i||_\infty \leq \sum_0^{j-1} ||g_i||_\infty$, $T_0(M^j + \sum_0^{j-1} h_i M^i + \lambda z)$ has closed range. Let $N_\theta = M^j + \theta(N - M^j)$, for any $0 \leq \theta \leq 1$. $T_0(N_\theta)$ has the same domain as $T_0(M^j)$, for any θ, by the preceding lemma. Define the index of any operator to be its nullity minus the deficiency of its range, provided that these numbers are finite. If Q is a regular ordinary differential expression with differentiable coefficients on $[a, \infty)$, and if $T_0(Q)$ has closed range, then index $T_0(Q) = $ deficiency of range $T_0(Q) = $ nullity $T_1(Q^+)$. A theorem on stability of the index under perturbation (theorem V.3.6, Goldberg (63)) states that if B is an operator such that $||Bf|| \leq a||f|| + b||T_0(Q)f||$ for all f in domain $T_0(Q)$, where a and b are non-negative real numbers such that $a + b||T_0(Q)^{-1}|| < ||T_0(Q)^{-1}||$, then index $T_0(Q) + B = $ index $T_0(Q)$. The operator $T_0(Q)^{-1}$ is here considered as an operator from range $T_0(Q)$ (a subspace of $L^2[a, \infty)$) into $L^2[a, \infty)$. $(T_0(Q))^{-1}$ is bounded, by the open mapping theorem, since the range of $T_0(Q)$ is closed.

It then follows that the index of $T_0(N_\theta + \alpha I)$ is constant for all θ in the interval $[0, 1]$. To see this, let θ_1 be any fixed point of $[0, 1]$, and we will show that the index is constant in a neighborhood of θ_1. The conclusion will then follow. But $N_\theta + \lambda z = N_{\theta_1} + \lambda z + (\theta - \theta_1)(N - M^j)$, and by the

preceding lemma, $T_0(N_\theta + \lambda z) = T_0(N_{\theta_1} + \lambda z) + (\theta - \theta_1)(N - M^j)$. Now, using

Lemma 2.6, and the previously mentioned theorem on the stability of the index,

we see that index $T_0(N_\theta + \lambda z)$ is constant for all θ in some neighborhood

of θ_1. Since this is true for all θ_1 in $[0, 1]$, then index $T_0(M^j + \lambda z) =$

index $T_0(N + \lambda z)$. Therefore nullity $T_1(M^j - \lambda z) =$ nullity $T_1(N^+ - \lambda z)$, as we

wished to show. The theorem is proved.

<u>Theorem 2.8</u>. Let $N = M^j + \sum_0^{j-1} c_i M^i$, where the c_i are complex numbers and M

is a formally symmetric, regular ordinary differential expression with C^∞

coefficients on $[a, \infty)$. Then $d(N) = d(M^j)$. Further, if the c_i are real,

$d^+(N) = d^+(M^j)$ and $d^-(N) = d^-(M^j)$.

Proof. Select $\lambda > \alpha$, where α is chosen as in Theorem 2.7. Then nullity

$T_1(N + \lambda z) =$ nullity $T_1(M^j + \lambda z)$, where $z = \pm(-1)^{1/2}$. However, by theorem 2.4.4,

$d(N + \lambda z) = d(N)$, and $d(M^j + \lambda z) = d(M^j)$. Furthermore, by theorem 2.4.1,

$d(N + \lambda z) =$ nullity $T_1(N + \lambda z) +$ nullity $T_1(N^+ - \lambda z)$, and

$d(M^j) =$ nullity $T_1(M^j + \lambda z) +$ nullity $T_1(M^j - \lambda z)$. Using Theorem 2.7 again, we

see that nullity $T_1(M^j - \lambda z) =$ nullity $T_1(N^+ - \lambda z)$. Thus $d(M^j) = d(N)$, and,

if $N = N^+$, $d^+(M^j) = d^+(N)$ and $d^-(M^j) = d^-(N)$. The theorem is proved.

<u>Remark</u>. Note that Theorem 2.8 proves the promised fact that any two polynomials

of the same degree in M have the same mean deficiency index.

<u>Definition 2.9</u>. Let M be a formally symmetric, regular ordinary differential

expression with C^∞ coefficients on $[a, \infty)$. Let $N = M^j + \sum_0^{j-1} C_k M^k$, where the

C_k are complex numbers. Then N is said to be <u>partially separated</u> if, for all

f in domain $T_1(N)$, $M^i f$ is in $L^2[a, \infty)$ for all $i \leq j$.

<u>Theorem 2.10</u>. Let M be a regular, formally symmetric, ordinary differential

expression with C^∞ coefficients on $[a, \infty)$. Assume that M is in the limit-point

condition. Let $N = M^j + \sum_0^{j-1} C_i M^i$, where the C_i are complex numbers. Then N

is limit-point if and only N is partially separated.

Proof. Lemma 2.6 shows that domain $T_0(N) =$ domain $T_0(M^j)$, and that $M^i f$ is in

L^2 for all $i \leq j$ when f is in domain $T_0(N)$. However, by Corollary 2.4.6, if N is in the limit-point condition, and f is in domain $T_1(N)$, there is a g in domain $T_0(N)$ such that $f - g$ is a C^∞ function supported in a compact interval $[a, r]$. Therefore, if N is in the limit-point condition, and f is in domain $T_1(N)$, $M^i f$ is in L^2 for all $i \leq j$.

Conversely, if N is partially separated, consider λ so large that $T_0(N + \lambda z)$ has closed range, where $z = \pm(-1)^{1/2}$. λ exists by Theorem 2.7 Now $N + \lambda z = p(M)$, where p is a complex polynomial. Let $p(M) = \pi_{i=1}^{j}(M - \lambda_i)$. Then domain $T_1(N + \lambda z) = $ domain $T_1(N)$, so $N + \lambda z$ is partially separated. Therefore $T_1(N + \lambda z) = \pi_{i=1}^{j} T_1(M - \lambda_i)$. This product commutes, and $T_1(N + \lambda z)$ is surjective (by 2.3.15), so $T_1(M - \lambda_i)$ is surjective for each i. Elementary linear algebra then shows that nullity $T_1(N + \lambda z) = \sum_{i=1}^{j}$ nullity $T_1(M - \lambda_i)$. Similarly $T_1(N^+ - \lambda z) = \sum_{i=1}^{j}$ nullity $T_1((M - \lambda_i)^+) = \sum_{i=1}^{j}$ nullity $T_1(M - \overline{\lambda}_i)$. Therefore $d(N + \lambda z) = j d(M) = d(N)$, as we see by using 2.4.4 and 2.4.2. This completes the proof.

3. The rule.

We now prove the following rule:

Theorem 3.1. Let M be any regular formally symmetric ordinary differential expression of order n with C^∞ coefficients on $[a, \infty)$. Take M^0 to be the identity operator, and $d(M^0) = 0$. Then, if $k \geq 1$, $d(M^k) - d(M^{k-1}) \leq n$. Furthermore, if $j \geq k \geq 1$, then

$$d(M^j) - d(M^{j-1}) \geq d(M^k) - d(M^{k-1}).$$

Remark. Thus, for example, if $n = 2$, and if $d(M^k)$ is known, and $d(M^k) \neq k$, then $d(M^r)$ is known for _every_ positive integer r. If $n > 2$, more possibilities occur, but they are still very restricted. Examples were given in chapter 5, sec. 5 to show that any sequence of deficiency indices not specifically prohibited by this rule actually does occur.

Proof of Theorem 3.1. Let i denote the imaginary unit, and I the identity transformation. Let R denote the non-symmetric expression $M + iI$, so that $R^+ = M - iI$.

It follows from the Schwartz inequality that $||Rf|| \geq ||f||$ for all f in $C_0^\infty(a, \infty)$, since $(Rf, f) = (Mf, f) + i(f, f)$, and (Mf, f) is real. It then follows that $||R^k f|| \geq ||f||$ for every k. Similarly, for every k, $||(R^+)^k f|| \geq ||f||$. Since minimal operators are the closures of their restrictions to $C_0^\infty(a, \infty)$, it follows that $T_0(R^k)$ and $T_0((R^+)^k)$ have closed range for every k. By Theorem 2.4.3,

$$2d(R^k) = \text{nullity } T_1(R^k) + \text{nullity } T_1((R^k)^+).$$

But $(R^k)^+ = (R^+)^k$, since the adjoint of the product is the product (in reverse order) of the adjoints. Therefore

(3.1) $$2d(R^k) = \text{nullity } T_1(R^k) + \text{nullity } T_1((R^+)^k).$$

However, by Theorem 2.8, $d(R^k) = d(M^k)$. Thus we have

(3.2) $$2d(M^k) = \text{nullity } T_1(R^k) + \text{nullity } T_1((R^+)^k).$$

We now prove that $d(M^k) - d(M^{k-1}) \leq n$. To see this, notice that $T_1(R^k)f = 0$ if and only if f is in L^2 and $R^k f = 0$. For this to happen, $R(R^{k-1}f)$ must vanish (although $R^{k-1}f$ need not be in L^2). Let $S(A)$ denote the null space of an operator A. Then

$$\text{nullity } T_1(R^k) - \text{nullity } T_1(R^{k-1}) = \text{dim. } V,$$

where V is the quotient space $S(T_1(R^k))/S(T_1(R^{k-1}))$. Since the null space of R has dimension less than or equal to n, $\dim V \leq n$. The same relationship holds for $(R^+)^k$, so

$$d(M^k) - d(M^{k-1}) \leq n.$$

Now we prove the second part of the rule. We consider the product operator $T_1(R) T_1(R^j)$. The domain of this operator is the set of all f in L such that $R^j f$ is in L and $R(R^j f)$ is also in L. Thus the domain of $T_1(R)(T_1(R^j))$ is contained in, but not necessarily equal to, the domain of $T_1(R^{j+1})$, and these two operators agree in the intersection of their domains.

From the quotient space argument used in the first part of the proof, we see that

(3.3) $\qquad$ nullity$(T_1(R)T_1(R^j)) \le$ nullity $T_1(R)$ + nullity $T_1(R^j)$.

However, equality actually holds in (3.3), because $T_0(R^j)$ has closed range and therefore, by Theorem 2.3.15, $T_1(R^j)$ is surjective. We have obtained

(3.4) $\qquad$ nullity $(T_1(R)T_1(R^j))$ = nullity $T_1(R)$ + nullity $T_1(R^j)$.

Now consider the following quotient space

$$Q_j = S(T_1(R^j))/S(T_1(R)T_1(R^{j-1})).$$

We wish to analyze the dimension of Q_j. First, we observe (using (3.4)) that

(3.5) $\qquad$ dim Q_j = nullity $T_1(R^j)$ - (nullity $T_1(R)$ + nullity $T_1(R^{j-1})$).

Now we examine this dimension in another way. Let $\{[f_i]\}$ be a basis for Q_j. Denote by $[f_i]$ the equivalence class corresponding to some function f_i. Each f_i is in $S(T_1(R^j))$, and no linear combination of the f_i is in $S(T_1(R)T_1(R^{j-1}))$. Thus each f_i is in L^2 and $T_1(R^j)f_i = 0$, but no linear combination of the $R^{j-1}f_i$ is in L^2.

Thus a basis $\{[g_m]\}$ of Q_k consists of a maximal set of equivalence classes such that g_m is in $S(T_1(R^k))$, and no linear combination of the $R^{k-1}g_m$ is in L^2. Since $T_1(R)$ is surjective, for each g_m there is a function h_m such that $T_1(R)h_m = g_m$. Pick one h_m for every g_m. (There are, of course, infinitely many choices for each h_m.) Clearly, h_m is in L^2. Since $R^k g_m = 0$, it is clear that $R^{k+1}h_m = 0$. Since no linear combination of the $R^{k-1}g_m$ is in L^2, it follows that no linear combination of the $R^k h_m$ is in L^2. Thus the equivalence classes $[h_m]$ are a linearly independent subset of Q_{k+1}. Therefore

(3.6) $\qquad$ dim $Q_{k+1} \ge$ dim Q_k.

From (3.6) and (3.5), we see that

(3.7) $\qquad$ nullity $T_1(R^{k+1})$ - nullity $T_1(R^k) \ge$ nullity $T_1(R^k)$ - nullity $T_1(R^{k-1})$.

The inequality (3.7) is also true for R^+. Therefore, it follows from (3.2) that the second part of the rule is proved.

4. Notes and comments.

J. Chaudhuri and W. N. Everitt [9] originated the study of the deficiency indices of powers of a differential expression, and gave an example of a second order M such that M is limit-point and M^2 is not limit-point.

The subject then received great impetus from a paper of Everitt and M. Giertz. In [48] they introduced the concept of partial separation, and gave a number of conditions on the coefficients of a second order M which guarantee that all powers are partially separated. They then use this to show that all powers are limit-point.

A. Zettl, in [137] generalizes some of the results of Everitt and Giertz to formally symmetric differential expressions of arbitrary order. He proved, among other things, that, if M is limit-point and p is a real polynomial, then p(M) is partially separated if and only if p(M) is limit-point. In [138] Zettl proved that, for $k = 2r$ and p a real polynomial, $d^+(p(M))$, $d^-(p(M)) \geq rd(M)$, and if $k = 2r + 1$, $d^+(p(M)) \geq (r + 1)d^+(M) + rd^-(M)$. In [138] he proved that equality holds if and only if p(M) is partially separated.

The theorems and techniques of proof of section 2 are due to R. M. Kauffman, [79]. The rule of section 3 is also due to Kauffman, [80].

CHAPTER VII

CONDITIONS ON THE COEFFICIENTS FOR ALL POWERS TO BE LIMIT-POINT.

1. Introduction.

Here we show that, with some minor modifications, the limit-point conditions developed in chapter 3 for second order expressions and in chapter four for $2n$-th order ones are sufficient not only for the expression M itself to be limit-point, as shown there, but also for any power M^k--and hence any polynomial in M--to be limit-point. The method of proof consists of establishing the partial separation property for M^k and then using Corollary 5.3.6. The technique for establishing that M^k is partially separated is quite lengthy and technical. Some of the lemmas from chapter 4 will be used as well as some new ones established.

Below, the notation, terminology and conventions used in chapter 4 will be followed. Since the second order case is somewhat different from the higher order one it is stated separately. However only the proof of the $2n$-th order case is presented in detail because the modifications needed for the proving the second order version are minor and straight-forward.

2. Sufficient conditions for M^k to be limit-point.

2.1 Theorem. Suppose the hypothesis of Theorem 4.2.1 hold. Then M^k is limit-point for any $k = 1,2,3,\ldots$ and hence by Theorem 6.2.8 all real polynomials in M are limit-point.

The conditions for the second order case are slightly different from those in chapter 3 so we state these in detail.

For $n = 1$ and putting $R^4/p_0 = w^2$ we get

2.2 Theorem. Suppose there exists a non-negative absolutely continuous function w and positive constants K_1, K_2, K_3, δ, and a such that on $[a, \infty)$

(i) $(1 + \delta)p_0 w'^2 - r_1 w^2 \leq K_1$

(ii) $w \leq K_2$

(iii) $w(t)p_0^{-\frac{1}{2}}(t)|h_1(t)| \leq K_2$, $\quad q_1(t) \geq h_1'(t)$,

(iv) $\int_a^\infty wp_0^{-\frac{1}{2}} = \infty$.

Then M^k is limit-point for any $k = 1,2,3,\ldots$. Consequently any real polynomial in M is limit-point.

<u>2.3 Remark</u>. Without condition (ii) these conditions correspond to those of Theorem 3.2.5 with (iii) here being a slight extension of (ii) there. The extra condition (ii) that w be bounded can be assumed without loss of generality if the two terms on the left in (i) are separately bounded as can be seen by specializing the proof of Lemma 4.3.1. This is so in many applications but it is not clear that this can be arranged to cover the example (4.2.1) when $n = 1$ for all values of α and β.

For other cases covered by Theorem 2.2 see chapter 3.

Before proceeding with the proof of Theorem 2.1 we state another lemma.

<u>2.3 Lemma</u>. Under the hypothesis of Lemma 4.3.2, given $\varepsilon > 0$ there exists a $K(\varepsilon) > 0$, independent of I, such that

$$(2.1) \qquad \int v^{4nj}(M^jh)^2 \leq \varepsilon \int v^{4n(j+1)}(M^{j+1}h)^2 + K(\varepsilon) \int v^{4n(j-1)}(M^{j-1}h)^2$$

for $j = 1,2,\ldots$.

Proof. Put $f = M^{j-1}h$, $g = Mf = M^jh$. Then $\int v^{4nj}fMg =$

$$\int v^{4nj}M^{j-1}hM^{j+1}h = \sum_{m=0}^{n}(-1)^m \int v^{4nj}f(p_{n-m}g^{(m)})^{(m)}$$

$$= \sum_{m=0}^{n}\sum_{\ell=0}^{m}\binom{m}{\ell}\int (v^{4nj})^{(m-\ell)}f^{(\ell)}p_{n-m}g^{(m)}$$

$$= \sum_{m=0}^{n}\int v^{4nj}f^{(m)}p_{n-m}g^{(m)}$$

$$\qquad + \sum_{m=1}^{n}\sum_{\ell=0}^{m-1}\binom{m}{\ell}\int (v^{4nj})^{(m-\ell)}f^{(\ell)}p_{n-m}g^{(m)}$$

$$= \sum_{m=0}^{n}(-1)^m \int v^{4nj}(p_{n-m}f^{(m)})^{(m)}g$$

$$\qquad + \sum_{m=1}^{n}\sum_{k=0}^{m-1}(-1)^m\binom{m}{k}\int (v^{4nj})^{(m-k)}(p_{n-m}f^{(m)})^{(k)}g$$

$$\qquad + \sum_{m=\ell}^{n}\sum_{\ell=0}^{m-1}\binom{m}{\ell}\int (v^{4nj})^{(m-\ell)}f^{(\ell)}p_{n-m}g^{(m)}$$

$$= \int v^{4nj}gMf$$

$$\qquad + \sum_{m=1}^{n}\sum_{k=0}^{m-1}\sum_{\ell=0}^{k}(-1)^{m+k}\binom{m}{k}\binom{k}{\ell}\int (v^{4nj})^{(m-\ell)}p_{n-m}f^{(m)}g^{(\ell)}$$

$$(2.2) \qquad + \sum_{m=1}^{n}\sum_{\ell=0}^{m-1}\binom{m}{\ell}\int (v^{4nj})^{(m-\ell)}p_{n-m}f^{(\ell)}g^{(m)}.$$

From (ii) in Theorem 4.2.1, for $1 \leq m \leq n$

$$\int q_{n-m} (v^{4nj})^{(m-\ell)} {}_f^{(m)} {}_g^{(\ell)}$$

$$= (-1)^{n-m} \int h_{n-m} \{ (v^{4nj})^{(m-\ell)} {}_f^{(m)} {}_g^{(\ell)} \}^{(n-m)}$$

$$= (-1)^{n-m} \sum_{k=0}^{n-m} \binom{n-m}{k} \sum_{s=0}^{k} \binom{k}{s} \int h_{n-m} (v^{4nj})^{(n-\ell-k)} {}_f^{(m+k-s)} {}_g^{(\ell+s)}$$

$$\leq K \sum_{k=0}^{m-n} \sum_{s=0}^{k} (F_{m+k-s} G_{\ell+s})^{\frac{1}{2}}$$

from (3.10) in chapter 4, part (v) of Theorem 4.2.1, and part (c) of Lemma 4.3.2 with $\alpha = 4n(j-1)$ in (3.3) and (3.4) of chapter 4. Also from (iv) of Theorem 4.2.1, for $m \geq 1$

$$\int r_{n-m} (v^{4nj})^{m-\ell} {}_f^{(m)} {}_g^{(\ell)} \leq K(F_m G_\ell)^{\frac{1}{2}} .$$

The last integral in (2.2) yields similar results and we get from (3.23) and (3.24) of chapter 4 with $\alpha = 4n(j-1)$

$$\int v^{4nj} fMg \geq \int v^{4nj} fMf - K \sum_{\substack{m,\ell=0 \\ m+\ell < 2n}}^{n} (F_m G)^{\frac{1}{2}}$$

$$= G_0 - K \sum_{m,\ell=0}^{n} \{ \varepsilon_1 G + (K/\varepsilon_1) F_m \}$$

$$\geq G_0 - K\varepsilon_1 [\varepsilon_2 \int v^{4n(j+1)} (M^2 f)^2 + K(\varepsilon_2) G_0]$$

$$- (K/\varepsilon_1)(\varepsilon_3 G_0 + K(\varepsilon_3) F_0).$$

Hence, by suitably choosing ε_1, ε_2 and ε_3, we get for any $\varepsilon > 0$

$$G_0 \leq \int v^{4nj} fMg + \varepsilon \int v^{4n(j+1)} (M^2 f)^2 + K(\varepsilon) \int v^{4n(j-1)} f^2$$

$$\leq \varepsilon_4 \int v^{4n(j+1)} (M^2 f)^2 + K(\varepsilon_4) \int v^{4n(j-1)} f^2$$

which is (2.1).

2.4 Lemma. Under the hypothesis of Lemma 2.3, given any $\varepsilon > 0$ there exists a $K(\varepsilon) > 0$, independent of I, such that

(2.3) $\qquad \int_I v^{4nj} (M^j h)^2 \leq \varepsilon \int_I v^{4nk} (M^k h)^2 + K(\varepsilon) \int_I h^2$

for $j = 1, 2, \ldots, k-1$.

Proof. The proof is by induction on k. The case $k = 2$ is Lemma 2.3. Suppose (2.3) holds for all integers up to k. Then, by Lemma 2.3 and the inductive hypothesis,

$$\int v^{4nk}({}_M{}^k h)^2 \le \varepsilon_1 \int v^{4n(k+1)}({}_M{}^{k+1}h)^2 + K(\varepsilon_1) \int v^{4n(k-1)}({}_M{}^{k-1}h)^2$$

$$\le \varepsilon_1 \int v^{4n(k+1)}({}_M{}^{k+1}h)^2$$

$$+ K(\varepsilon_1)\{\varepsilon_2 \int v^{4nk}({}_M{}^k h)^2 + K(\varepsilon_2) \int h^2\}$$

which gives for arbitrary $\varepsilon > 0$, by a suitable choice of ε_1 and ε_2,

(2.4) $\qquad \int v^{4nk}({}_M{}^k h)^2 \le \varepsilon \int v^{4n(k+1)}({}_M{}^{k+1}h)^2 + K(\varepsilon) \int h^2$.

Also if $s < k$, we get from the induction hypothesis and (2.4)

$$\int v^{4ns}({}_M{}^s h)^2 \le \varepsilon_1 \int v^{4nk}({}_M{}^k h)^2 + K(\varepsilon_1) \int h^2$$

$$\le \varepsilon \int v^{4n(k+1)}({}_M{}^{k+1}h)^2 + K(\varepsilon) \int h^2$$.

This completes the proof.

Proof of Theorem 2.1. We show that it follows from Lemma 2.4 that M^k is partially

separated for $k = 2,3,4,\ldots$. The result then follows from Corollary 5.3.6.

From (x) of Theorem 4.2.1 and $p_0 > 0$ it follows that for some $b > a$

$R^{4n-2}(b) > 0$ and hence, since R is continuous, there is a $\delta > 0$ such that

$R^{4n-2}/p_0 > 0$ on $[b, b+\delta]$. Define

$$\theta(t) = \int_b^t R^{4n-2}/p_0, \quad t \ge b$$

$$v(t) = \begin{cases} 1 - \exp[\theta(t) - \theta(T), & b + \delta \le t \le T \\ 0 & , & t \ge T \end{cases}$$

and in $[b, b+\delta)$ choose v such that it vanishes in a right neighborhood of b,

$0 \le v(t) \le 1$ and v has n continuous derivatives in $[b, b+\delta]$. Then from (ix),

the boundedness of R^{4n}/p_0 which we can assume by Lemma 4.3.1, we have

(2.5) $\qquad v^{(j)} = 0(R^{4n-2j}/p_0), \quad j = 1,2,\ldots,n.$

Now we choose T such that $\theta(T) > \log 2$ and we choose x such that

$\theta(x) = \theta(T) - \log 2$. Then

(2.6) $\qquad v(t) \ge 1/2, \quad b + \delta \le t \le x.$

Now from the definition of v and Lemma 2.4 we get

$$(1/2)^{4nj} \int_{b+\delta}^{x} (M^j h)^2 \leq \int_{b+\delta}^{T} v^{4nj} (M^j h)^2$$

$$\leq K \int_{0}^{\infty} \{ (M^k h)^2 + h^2 \} \; .$$

Since $x \to \infty$ as $T \to \infty$ we can conclude that $M^j h$ is in $L^2(0, \infty)$ for $j = 1, 2, \ldots, k-1$ and that M^k is partially separated. Thus the proof of Theorem 2.1 is complete.

Now we give a couple of applications of Theorem 2.1.

2.5 Corollary. Let $p_0 = 1$ and suppose there exists a sequence of disjoint intervals I_m of length ℓ_m such that in each I_m

(i) $\quad p_i = h_{i,m}^{(i)}, \quad i = 1, 2, \ldots, n-1$

(ii) $\quad p_n \geq h_{n,m}^{(n)}$

(iii) $\quad \ell_m^{n-i} |h_{n-i,m}| \leq K, \quad i = 0, 1, \ldots, n-1$

(iv) $\quad \sum_{m=1}^{\infty} \ell_m^{2n} = \infty \; .$

Then M and all its powers M^k are limit-point $k = 2, 3, \ldots$.

Proof. Let $I_m = (a_m, b_m)$ and take $r_i = 0, \quad i = 1, \ldots, n$ in Theorem 2.1. Define

$$R(t) = \begin{cases} (t - a_m)^{\frac{1}{2}}, & a_m \leq t \leq (a_m + b_m)/2 \\ (b_m - t)^{\frac{1}{2}}, & (a_m + b_m)/2 \leq t \leq b_m \\ 0 & \text{otherwise.} \end{cases}$$

As a very special case of Corollary 2.5 we have

2.6 Corollary. The differential expression M given by (1.1) of chapter 4 and all its powers (and hence all polynomials in M) are limit-point if $p_0 = 1$ and there exists a sequence of intervals I_m of length

$$\ell_m \geq \delta > 0, \quad m = 1, 2, 3, 4, \ldots \text{ and some } \delta \text{ and there exist positive}$$

constants $K_j, \; j = 1, \ldots, n$ such that on these intervals I_m

(i) $\quad |p_j| \leq K_j, \quad j = 1, \ldots, n-1$

(ii) $\quad p_n \geq -K_n \; .$

An interesting feature of Corollaries 2.5 and 2.6 is the fact that no conditions at all--other than the basic ones given by (1.2) of chapter 4--are imposed on the coefficients p_j outside the intervals I_m. In particular these conditions permit wildly oscillatory behavior of the coefficients p_j. This feature is illustrated with the examples (2.4) of chapter 4. The proof of Corollary 4.2.4 together with Theorem 2.1 (or Corollary 2.5) shows that M and all its powers are limit-point under the conditions given there. We state this as

2.7 Corollary. Let

$$My = y^{(2n)} + t^\alpha \sin t^\beta y .$$

Then M^k is limit-point for all $k = 1,2,3,\ldots$ under any one of the following conditions:

(i) $\alpha \leq 2n/(2n-1)$, all β

(ii) $\beta \leq 2n/(2n-1)$, all $\alpha \geq 0$

(iii) $\alpha \leq n\beta - 2n(n-1)/(2n-1)$.

We summarize these results in a diagram

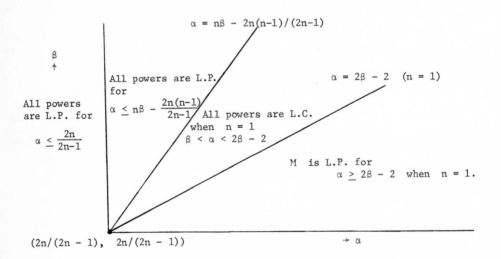

All powers are L.P. for $\beta \leq 2n/(2n-1)$

Question. Let $n = 2$. For what values of α, β, if any, is $d(M) = 3$? Are there any values of α and β such that $d(M) = 4$ i.e. M is limit-circle. An affirmative answer to the second question would be particularly interesting as there are presently no known examples of functions q such that $y^{(4)} + qy$ is limit-circle.

Another interesting feature of these conditions for all powers M^k of M to be limit-point is that no assumptions whatever are needed on the derivatives of the coefficients. When one actually computes, say M, in the classical way i.e. no quasi-derivatives one finds, of course, (see chapter 5, section 1) that the coefficients of M involve the derivatives of the coefficients p_j of M. So it does seem surprising that no conditions are needed on these derivatives to get information about $d(M^2)$, say.

In view of the importance of the second order case we state some consequences of Theorem 2.2.

2.8 Corollary. Let $n = 1$, i.e. M is given by

$$My = -(p_0 y')' + p_1 y.$$

Let I_m be a sequence of disjoint intervals of length ℓ_m and let $p_m = \inf$ over I_m of p_0, $P_m = \int_{I_m} p_0^{-\frac{1}{2}}$ and

(a) $\int_\alpha^\beta p_1 \geq -k p_m^{\frac{1}{2}}/P_m$ for all $[\alpha, \beta] \subseteq I_m$,

(b) $\sum_{m=1}^\infty P_m^2 = \infty$.

Then M and all its powers M^k and hence all real polynomials in M by 6.2.8 are limit-point.

Proof. In Theorem 2.2 let $r_1 = 0$ and $w(t) = \sum_{m=1}^\infty w_m(t)$ where support $w_m = I_m = [a_m, b_m]$ and

$$w_m(t) = \begin{cases} \int_a^t p_0^{-\frac{1}{2}}, & a_m \leq t \leq c_m \\ \int_t^{b_m} p_0^{-\frac{1}{2}}, & c_m \leq t \leq b_m \end{cases}$$

c_m being defined by $\int_{a_m}^{c_m} p_0^{-\frac{1}{2}} = \int_{c_m}^{b_m} p_0^{-\frac{1}{2}} = P_m/2$. It follows that $|p_0 w'^2| \leq 1$ and hence, since $r_1 = 0$, that (i) of Theorem 2.2 holds.

In [2], Atkinson proved that (a) implies the existence of a function Q_m which is absolutely continuous on I_m and is such that

$$-p_1(t) \le Q_m'(t), \quad |Q_m(t)| \le Kp_m^{\frac{1}{2}}/P_m, \quad t \text{ in } I_m.$$

This implies (iii) of Theorem 2.2 and the rest follows easily.

2.9 Corollary. Let M be as in Corollary 2.8. Suppose there exist disjoint intervals I_m, $m = 1,2,3,\ldots$ and positive numbers v_m such that, with $P_m = \int_{I_m} p_0^{-\frac{1}{2}}$ and $p_1^- = \min(0, p_1)$,

(a) $v_m P_m^2 \ge 1$

(b) $\sum_{m=1}^{\infty} v_m^{-1} = \infty$

(c) $\int_{I_m} p_1^- \ge -Kv_m^2 P_m^3 p^{\frac{1}{2}}$.

Then M^k is limit-point for all $k = 1,2,3,\ldots$.

Proof. In fact Corollary 2.8 with p_1 in (a) replaced by p_1^- is equivalent to Corollary 2.9.

Suppose the conditions of Corollary 2.9 hold and divide I_m into $N(m)$ disjoint intervals $J_{m,k}$ such that

$$N(m) \int_{J_{m,k}} p_0^{-\frac{1}{2}} = P_m, \quad k = 1,2,\ldots,N(m)$$

and

$$N(m) = [v_m P_m^2] + 1.$$

(Here $[x]$ denotes the greater integer $\le x$.)

Let

$$S_m = \{k \mid 1 \le k \le N(m), \int_{I_{m,k}} -p_1^- \le 4Kv_m^2 P_m^3 p_m^{\frac{1}{2}}/N(m) \}$$

where K is as in (c). Then S_m has at least $N(m)/2$ members, for otherwise

$$\int_{I_m}(-p_1^-) \ge \sum_k \ntrianglelefteq S_m \int_{J_{m,k}}(-p_1^-) \ge 2Kv_m^2 P_m^3 p_m^{\frac{1}{2}}$$

contrary to (c). Let $P_{m,k} = \inf.$ over $J_{m,k}$ of $p_0(t)$ and $P_{m,k}^- = \int_{J_{m,k}} p_0^-$.

Then $P_m = P_{m,k} N(m) \ge P_{mk} v_m P_m^2$ and so for k in S_m

$$\int_{J_{m,k}}(-p_1^-) \le 4Kp_m^{\frac{1}{2}} v_m P_m \le 4Kp_m^{\frac{1}{2}}/P_{m,k} .$$

Also

$$\sum_m \sum_{k \,\in\, S_m} P_{mk}^2 \geq \sum_m N(m) P_m^{\,2} / 2N^2(m)$$

$$\geq K \sum_m v_m^{-1} = \infty \ .$$

Thus Corollary 2.8 implies 2.9. The converse is obtained by putting $v_m = 1/P_m^{\,4}$.

3. Notes and comments.

The results of this chapter are due to Evans and Zettl [35, 36, 37] and are taken directly from these papers. Special cases of Theorem 2.2 were previously obtained, by different methods, by Chaudhuri and Everitt, [9], Everitt and Giertz [47-53]. No conditions on the coefficients for powers of higher order expressions to be limit-point seem to have been known prior to the publication of Evans-Zettl [36, 37]. The authors are greatly indebted to W. D. Evans for permission to use these results prior to the appearance of these papers.

Some results independent of Theorem 2.2 for powers of second order nonoscillatory expressions were obtained by Read [109, 111]. In contrast to Theorem 2.2, most of these assert only that some of the powers are limit-point. It is, in fact, common for such expressions to have some powers limit-point, but to fail to have all powers limit-point. It is shown in [109] and [111] that the results do find all powers which are limit-point for some simple classes of nonoscillatory expressions. One of these results (Theorem 7.1 of [111]) is as follows. (Recall from Proposition 3.4.8 that $-(py')' + qy = 0$ has a positive solution on $[a, \infty)$ if and only if $q = q_1 + q_2$ with $p(t)q_1(t) \geq (c + \int_a^t q_2)^2$ for all $t \geq a$.)

Theorem. If there are constants $\delta > 0$, k, and

$$K > [(n - \tfrac{1}{2})^2 \, (1 + \delta)/\delta]^{\frac{1}{2}}$$

and a decomposition $q = q_1 + q_2$ such that

(i) $p(t)(q_1(t) + k) \geq (1 + \delta)(\int_a^t q_2)^2$ for all $t \geq a$,

(ii) $\int_a^t [(q_1 + k)/p]^{\frac{1}{2}} \geq K \log p(t)$ for all t in a set of infinite measure,

then every polynomial in $-(py')' + qy$ of degree at most n is limit-point.

REFERENCES

1. N. I. Akhiezer and I. M. Glazman, "Theory of Linear Operators in Hilbert Space," v. II, Ungar, New-York, 1963.

2. F. V. Atkinson, "Limit-n criteria of integral type," Proc. Roy. Soc. Edinburgh (A), 73, 11, 1975, 167-198.

3. F. V. Atkinson, M. S. P. Eastham and J. B. McLeod, "The limit-point, limit-circle nature of rapidly oscillating potentials," Proc. Roy. Soc. Edinburgh (A), (to appear).

4. F. V. Atkinson and W. D. Evans, "On solutions of a differential equation which are not of integrable square," Math. Z. 127 (1972) 323-332.

5. F. V. Atkinson, W. N. Everitt, and K. S. Ong, "On the m-coefficient of Weyl for a differential equation with an indefinite weight function," Proc. London Math. Soc. (3) 29 (1974) 368-384.

6. J. S. Bradley, "Comparison theorems for the square integrability of solutions of $(r(t)y') + q(t)y = f(t,y)$", Glasgow Mathematical Journal, Vol. 13, part 1, (1972), 75-79.

7. I. Brinck, "Self-adjointness and spectra of Sturm-Liouville operators," Math. Scand. 7(1959) 219-239.

8. B. M. Brown and W. D. Evans, "On the limit-point and strong limit-point classification of 2n-th order differential expressions with wildly oscillating coefficients," Math. Z. 134 (1973) 351-368.

9. J. Chaudhuri and W. N. Everitt, "On the square of a formally self-adjoint differential expression," J. London Math. Soc. (2), 1 (1969) 661-673.

10. W. A. Coppel, "Disconjugacy", Lecture Notes in Mathematics no. 220, Springer Verlag, Berlin, 1970.

11. _____, "Stability and asymptotic behavior of differential equations," D. C. Heath, Boston, 1965.

12. A. Devinatz, "The deficiency index of ordinary self-adjoint differential operators," Pac. J. Math. 16 (1966), 243-257.

13. _____, "The deficiency index of a certain class of ordinary self-adjoint differential operators," Advances in Math. 8 (1972) 434-473.

14. _____, "The deficiency index of certain fourth order differential operators," Quart. J. Math. Oxford (2), 23 (1972) 267-286.

15. _____, "Positive definite fourth-order differential operators," J. London Math. Soc. (2), 6 (1973), 412-416.

16. _____, "The deficiency index problem for ordinary self-adjoint differential operators," Bull. Amer. Math. Soc. 79 (1973) 1109-1127.

17. _____, "On limit-2 fourth-order differential operators," J. London Math. Soc. (2) 6 (1973).

18. _____, "An asymptotic theorem for systems of linear differential equations," Trans. Amer. Math. Soc., 160 (1971) 353-363.

19. _____, "On a deficiency index theorem of W. N. Everitt," Lecture Notes in Mathematics 415, Springer Verlag, (1974) 103-108.

20. A. Devinatz and J. Kaplan, "Asymptotic estimates for solutions of linear systems of ordinary differential equations having multiple characteristic roots," Indiana Univ. Math. J. 22 (1972) 355-366.

21. N. Dunford and J.T.Schwartz, "Linear operators," II, Interscience, New York, 1963.

22. M. S. P. Eastham, "On the L^2 classification of fourth-order differential equations," J. London Math. Soc., (2) 3 (1971) 297-300.

23. _____, "The limit-2 case of fourth-order differential equations," Quart. J. Math. (2) 22 (1971) 131-134.

24. _____, "On a limit-point method of Hartman," Bull. London Math. Soc. 4 (1972) 340-344.

25. _____, "Limit-circle differential expressions of the second order with an oscillating coefficient," Quart. J. Math. Oxford (2) 24 (1973).

26. _____, "Semi-bounded second-order differential operators," Proc. Roy. Soc. Edinburgh (A), 72, 2, (1973) 9-16.

27. _____, "The limit-3 case of self-adjoint differential expressions of the fourth order with oscillating coefficients," J. London Math. Soc. (2), 8 (1974), 427-437.

28. M. S. P. Eastham and A. A. El-Deberky, "The spectrum of differential operators with large coefficients," J. London Math. Soc. (2), 2 (1970), 257-266.

29. M. S. P. Eastham and M. L. Thompson, "On the limit-point, limit-circle classification of second-order ordinary differential expressions," Quart. J. Math. Oxford Ser. (2), 2 (1973) 531-535.

30. M. S. P. Eastham and A. Zettl, "Second-order differential expressions whose squares are limit-3," Proc. Roy. Soc. Edinburgh, A (to appear).

31. M. S. P. Eastham, W. D. Evans and J. B. McLeod, "The essential self-adjointness of Schrödinger-type operators, M.R.C. technical report #1457 (1974).

32. W. D. Evans, "On the limit-point, limit-circle classification of a second-order differential equation with a complex coefficient," J. London Math. Soc. (2), 4 (1971), 245-256.

33. _____, "On non-integrable square solutions of a fourth-order differentiable equation and the limit-2 classification," J. London Math. Soc. (2), 7 (1973), 343-354.

34. _____, "On limit-point and Dirichlet-type results for second-order differential expressions," Proceedings of the 1976 Dundee Conference on Differential Equations, Lecture Notes in Mathematics, Springer-Verlag (to appear).

35. W. D. Evans and A. Zettl, "On the deficiency indices of powers of real 2nth-order symmetric differential expressions," J. London Math. Soc. (2), 13 (1976), 543-556.

36. _____, "Levinson's limit-point criterion and powers," J. Math. Anal. and Appl. (to appear).

37. _____, "Interval limit-point criteria for differential expressions and their powers," J. London Math. Soc. (to appear).

38. W. N. Everitt, "Integrable-square solutions of ordinary differential equations," Quart. J. Math. Oxford (2) 10 (1959) 145-155; II 13 (1962) 217-220; III, 14 (1963) 170-180.

39. _____, "Some positive definite differential operators," J. London Math. Soc. 43 (1968) 465-473.

40. _____, "On the limit-point classification of fourth-order differential operators," J. London Math. Soc., 44 (1969) 273-281.

41. _____, "On the limit-circle classification of second order differential expressions," Quart. J. Math. Oxford (2) 23 (1972) 193-196.

42. _____, "Integrable-square solutions of ordinary differential equations,"

43. _____, "Fourth order singular differential equations," Math. Annalen 149, (1963) 320-340.

44. _____, "On an eigenfunction expansion for a fourth-order singular differential equation," Quart. J. Math. (2) 20 (1969) 195-213. This paper is joint with J. Chaudhuri.

45. _____, "Singular differential equations I: The even order case," Math. Annalen 156 (1964) 9-24.

46. _____, "Singular differential equations II: Some self-adjoint even order cases," Quart. J. Math. (2) 18 (1967), 13-32.

47. W. N. Everitt and M. Giertz, "Some properties of the domains of certain differential operators," Proc. London Math. Soc. 23 (1971) 301-324.

48. _____, "On some properties of the powers of a formally self-adjoint differential expression," Proc. London Math. Soc. 24 (1972) 149-170.

49. _____, "On some properties of the domains of powers of certain differential operators," Proc. London Math. Soc. 24 (1972) 756-768.

50. _____, "A Dirichlet type result for ordinary differential operators," Math. Ann. 203 (1973) 119-128.

51. _____, "On the integrable-square classification of ordinary symmetric differential expressions," J. London Math. Soc. 10 (1975) 417-426.

52. _____, "On the deficiency indices of powers of formally symmetric differential expressions," Lecture Notes in Mathematics no. 448, 167-181, Springer-Verlag, Berlin-Heidelberg-New York, 1975.

53. _____, "A critical class of examples concerning the integrable-square classification of ordinary differential expressions," Proc. Roy. Soc. Edinburgh (A) 74 (1976) 285-298.

54. W. N. Everitt, M. Giertz, and J. B. McLeod, "On the strong and weak limit-point classification of second-order differential expressions," Proc. London Math. Soc. 29 (1974) 142-158.

55. W. N. Everitt, M. Giertz, and J. Weidmann, "Some remarks on a separation and limit-point criterion of second-order ordinary differential expressions," Math. Ann. 200 (1973) 335-346.

56. W. N. Everitt, D. B. Hinton, and J. S. W. Wong, "On the strong limit-n classification of linear ordinary differential expressions of order 2n," Proc. London Math. Soc. 29 (1974) 351-367.

57. W. N. Everitt and V. K. Kumar, "On the Titchmarsh-Weyl theory of ordinary symmetric differential expressions I: The general theory," Nieuw Archief voor Wiskuude (3), XXIV (1976), 1-48.

58. W. N. Everitt and A. Zettl, "The number of integrable-square solutions of products of differential expressions," Proc. Roy. Soc. Edinburgh A, (to appear)

59. M. V. Fedorjuk, "Asymptotic methods in the theory of one-dimensional differential equations," Trudy Moskov. Mat. Obsc. 15 (1966) 295-346. Transl. Moscow Math. Soc. 1966, 333-386.

60. _____, "Asymptotic properties of the solutions of ordinary n^{th} order linear differential equations," Differencial'nge Uravneniya 2 (1966) 492-501--Differential Equations 2 (1966) 250-258.

61. I. M. Glazman, "On the deficiency indices of differential operators," Doklady. Akad. Nauk SSSR, 644 (1949) 151-154. (Russian)

62. _____, "On the theory of singular differential operators," Uspehi. Math. Nauk. 40 (1950) 102-135; English transl. Amer. Math. Soc. Transl. (1) 4 (1962) 331-372.

63. S. Goldberg, "Unbounded linear operators," McGraw Hill, New York, 1966.

64. I. Halperin, "Closures and adjoints of linear differential operators," Ann. of Math. 38 (1937) 880-919.

65. S. G. Halvorsen, "On the quadratic integrability of solutions of $x'' + fx = 0$, Math. Scand. 14(1964), 111-119.

66. P. Hartman, "On the number of L^2-solutions of $x'' + q(t)x = 0$," Amer. J. Math. 73(1951) 635-645.

67. _____, "On differential equations with non-oscillatory eigenfunctions," Duke Math. J. 15 (1948), 697-709.

68. P. Hartman and A. Wintner, "A criterion for the non-degeneracy of the wave equation," Amer. J. Math. 7 (1949), 206-213.

69. _____, "Criteria of non-degeneracy of the wave equation," Amer. J. Math. 70 (1948), 295-308.

70. D. B. Hinton, "Asymptotic behavior of the solutions of $(ry^{(m)})^{(k)} \pm qy = 0$," J. Differential Equations, 4 (1968) 590-596.

71. _____, "Limit-point criteria for differential equations," Can. J. Math. (to appear).

72. _____, "Solutions of $(ry^{(n)})^{(n)} + qy = 0$ of class $L_p(0, \infty)$," Proc. Amer. Math. Soc. 32 (1972) 134-138.

73. _____, "Limit-point criteria for positive definite fourth-order differential operators," Quart. J. Math. (to appear).

74. R. S. Ismagilov, "Conditions for self-adjointness of differential equations of higher order," Soviet Math. 3 (1962) 279-283.

75. _____, "On the self-adjointness of the Sturm-Liouville operator," Uspehi. Mat. Nauk. 18 (1963) 161-166.

76. T. Kato, "Perturbation theory for linear operators," Springer-Verlag, New York 1966.

77. H. Kalf, "Remarks on some Dirichlet-type results for semi-bounded Sturm-Liouville operators," Math. Ann. 210 (1974) 197-205.

78. B. Karlsson, "Generalization of a theorem of Everitt," J. London Math. Soc. (2) 9(1974), 131-141.

79. R. M. Kauffman, "Polynomials and the limit-point condition," Trans. Amer. Math. Soc. 201 (1975) 347–366.

80. _____, "A rule relating the deficiency index of L^j to that of L^k," Proc. Roy. Soc. Edinburgh (A) 74 (1976) 115–118.

81. _____, "Disconjugacy and the limit-point condition for fourth order operators,"

82. _____, "On the limit-n classification of ordinary differential operators with positive coefficients," to appear in Proceedings of the 1976 Dundee conference on differential equations. Lecture Notes in Mathematics, Springer-Verlag, 1976.

83. I. Knowles, "On second order differential operators of limit-circle type," Lecture Notes in Mathematics, no. 415, Springer-Verlag, 1974, 184–187.

84. _____, "Note on a limit-point criterion," Proc. Amer. Math. Soc. 41 (1973) 117–119.

85. _____, "A limit-point criterion for a second-order linear differential operator," J. London Math. Soc. 8 (1974) 719–727.

86. V. I. Kogan and F. S. Rofe-Beketov, "On the question of deficiency indices of of differential operators with complex coefficients," Proc. Roy. Soc. Edinburgh A72 (1973/1974).

87. K. Kodaira, "On ordinary differential equations of any even order and the corresponding eigenfunction expansions," Amer. J. Math. 72 (1950) 502–544.

88. K. V. Kumar, "A criterion for a formally symmetric fourth order differential expression to be in the limit-2 case at ∞," J. London Math. Soc. 8 (1974) 134–138.

89. _____, "The strong limit-2 case of fourth-order differential equations," Proc. Roy. Soc. Edinburgh (A), 71, 26 (1972/1973) 297–304.

90. N. P. Kupcov, "Conditions of non-self adjointness of a second order linear differential operator," Dokl. Akad. Nauk. SSSR 138 (1961) 767-770.

91. H. Kurss, "A limit-point criterion for non-oscillatory Sturm-Liouville differential operators," Proc. Amer. Math. Soc. 18 (1967) 445-449.

92. M. K. Kwong, "L^p-perturbations of second order linear differential equations," Math. Am. 215 (1975) 23-34.

93. _____, "On boundedness of solutions of second order differential equations in the limit-circle case," Proc. Amer. Math. Soc. 52 (1975) 242-246.

94. _____, "Note on a strong limit-point condition of second order differential expressions," Quart. J. Math. (to appear).

95. N. Levinson, "Criteria for the limit-point case for second-order linear differential operators," Casopis Pest. Mat. Fys. 74 (1949) 17-20.

96. J. B. McLeod, "The number of integrable-square solutions of ordinary differential equations," Quart. J. Math. Oxford 17 (1966) 285-290.

97. M. A. Naimark, "Linear differential operators," GITTL, Moscow 1954; English transl. Ungar, New York, part I 1967; part II 1968.

98. F. A. Neimark, "On the deficiency index of differential operators," Uspehi. Mat. Nauk. 17 (1962) 157-163.

99. M. Otelbaev, "On summability with a weight of a solution of the Sturm-Liouville equation," Mat. Zametki 16 (1974) 969-980.

100. S. A. Orlov, "On the deficiency index of differential operators," Dokl. Akad. Nauk. SSSR, 92 (1953) 483-486.

101. W. T. Patula, "A limit-point criterion for a fourth-order differential operator," J. London Math. Soc. 8 (1974) 217-225.

102. W. T. Patula and P. Waltman, "Limit-point classification of second-order linear differential equations," J. London Math. Soc. 8 (1974) 209-216.

103. W. T. Patula and J. S. W. Wong, "An L^p-analogue of the Weyl alternative," Math. Ann. 197 (1972) 9-28.

104. A. Pleijel, "Some remarks about the limit-point and limit-circle theory," Arkiv. Mat. 7 (1969) 543-550.

105. _____, "Complementary remarks about the limit-point and limit-circle theory," ibid. 8 (1971) 45-47.

106. C. R. Putnam, "On the spectra of certain boundary value problems," Amer. J. Math. 71 (1949), 109-111.

107. I. M. Rapaport, "On the asymptotic behavior of solutions of linear differential equations," Dokl. Akad. Nauk. SSSR, 78 (1951) 1097-1100. (Russian)

108. _____, "On a singular boundary value problem for ordinary differential equations," Dokl. Akad. Nauk. SSSR, 79 (1951) 21-24. (Russian)

109. T. T. Read, "On the limit-point condition for polynomials in a second order differential expression," J. London Math. Soc. 10 (1975) 357-366.

110. _____, "Sequences of deficiency indices," Proc. Roy. Soc. Edinburgh (A) 74 (1976) 157-164.

111. _____, "Limit-point criteria for polynomials in a non-oscillatory expression," Proc. Roy. Soc. Edinburgh, (A) 76 (1976) 13-29·

112. _____, "A limit-point criterion for expressions with oscillatory coefficients," Pac. J. Math. 66 (1976) 243-255.

113. _____, "A limit-point criterion for $-(py')' + qy$," Lecture Notes in Mathematics, Springer-Verlag, 1976.

114. _____, "A limit-point criterion for expressions with intermittently positive coefficients," J. London Math. Soc. (to appear)

115. D. B. Sears, "Note on the uniqueness of the Green's functions associated with certain differential expressions," Can. J. Math. 2 (1950), 314-325.

116. D. Shin, "On the solutions in $L^2(0, \infty)$ of the self-adjoint differential equations $u^{[n]} = \ell(u)$, $I(\ell) \neq 0$," Doklad. Akad. Nauk. SSSR, 18, (1938) 519-522 (Russian).

117. _____, "On quasi-differential operators in Hilbert space," Doklad. Akad. Nauk. SSSR, (1938) 18, 523-526. (Russian)

118. _____, "On the solutions of a linear quasi-differential equation of the nth order," Mat. Sb. 7 (49), (1940) 479-532 (Russian).

119. _____, "On quasi-differential operators in Hilbert space," Mat. Sb. 13 (55), (1943), 39-70 (Russian).

120. _____, "Existence theorems for the quasi-differential equation of the nth order," Dokl. Akad. Nauk. SSSR, 8 (1938).

121. E. C. Titchmarsh, "Eigenfunction expansions associated with second-order differential equations: part I," Oxford University Press, 1962.

122. K. Unsworth, "Asymptotic expansions and deficiency indices associated with third order self-adjoint differential operators," J. Math Oxford (2) 24 (1973),

123. P. W. Walker, "Asymptotics of the solutions to $[(ry'')' - py']' + qy = y$," J. Differential Equations, 9 (1971) 108-132.

124. _____, "Deficiency indices of fourth order singular differential operators," J. Differential Equations, 9 (1971) 133-140.

125. _____, "A vector-matrix formulation for formally symmetric ordinary differential equations with Applications to solutions of integrable-square," J. London Math. Soc. 9 (1974) 151-159.

126. _____, "Asymptotics for a class of weighted eigenvalue problems,"
Pac. J. Math. 40 (1972) 501–510.

127. _____, "Asymptotics for a class of fourth-order differential equations,"
J. Differential Equations, II (1972) 321–334.

128. _____, "Weighted singular differential operators in the limit-circle
case," J. London Math. Soc., 4 (1972) 741–744.

129. J. Walter, "Bemerkungen zu dem Grenzpunktfallkriterium von N. Levinson,"
Math. Z. 105 (1968) 345–350.

130. H. Weyl, "Über gewöhnliche Differentialgleichungen mit Singularitäten und
die zugehörigen Entwicklungen willkürlicher Funktionen," Math. Ann. 68 (1910)
220–269.

131. W. Windau, "Über lineare Differentialgleichungen vierter Ordnung mit
Singuläritaten und die zugenhörige Darstellung willkürlicher Funktionen."
Math. Ann. 83 (1921), 256–279.

132. J. S. W. Wong, "On L^2-solutions of linear ordinary differential equations,"
Duke Math. J. 38 (1971) 93–97.

133. J. S. W. Wong and A. Zettl, "On the limit-point classification of second
order differential equations," Math. Z. 132 (1973) 297–304.

134. A. D. Wood, "Deficiency indices of some fourth-order differential operators,"
J. London Math. Soc. 3 (1971) 96–100.

135. A. Zettl, "A note on square-integrable solutions of linear differential
equations," Proc. Amer. Math. Soc. 21 (1969) 671–672.

136. _____, "Square integrable solutions of $Ly = f(t, y)$," Proc. Amer.
Math. Soc. 26 (1970) 635–639.

137. _____, "The limit-point and limit-circle cases for polynomials in a
differential operator," Proc. Roy. Soc. Edinburgh, (A) 72 (1974) 219–224.

138. _____, "Deficiency indices of polynomials in symmetric differential expressions," Lecture Notes in Mathematics, no. 415 Springer-Verlag 293-301. II, Proc. Roy. Soc. Edinburgh 73A 20(1974/75) 301-306.

139. _____, "Formally self-adjoint quasi-differential operators," Rocky Mountain J. Math. 5 (1975) 453-474.

140. _____, "Perturbation of the limit-circle case," Quart. J. Math. Oxford (1975)

141. _____, "Perturbation theory of deficiency indices of differential operators," J. London Math. Soc. (2), 12 (1976), 405-412.

142. _____, "An algorithm for the construction of limit-circle expressions," Proc. Roy. Soc. Edinburgh (A) 75 (1975/76) 1-3.

143. _____, "Limit-point conditions for powers," Lecture Notes in Mathematics, Springer-Verlag 564, 540-550. See also Addendum to "Limit Point Conditions for Powers" by W. D. Evans and A. Zettl ibid, 550-551.

144. _____, "Powers of symmetric differential expressions without smoothness assumptions," Quaestiones Mathematicae, 1 (1976), 83-94.

Subject Index

Vol. 460: O. Loos, Jordan Pairs. XVI, 218 pages. 1975.

Vol. 461: Computational Mechanics. Proceedings 1974. Edited by J. T. Oden. VII, 328 pages. 1975.

Vol. 462: P. Gérardin, Construction de Séries Discrètes p-adiques. »Sur les séries discrètes non ramifiées des groupes réductifs déployés p-adiques«. III, 180 pages. 1975.

Vol. 463: H.-H. Kuo, Gaussian Measures in Banach Spaces. VI, 224 pages. 1975.

Vol. 464: C. Rockland, Hypoellipticity and Eigenvalue Asymptotics. III, 171 pages. 1975.

Vol. 465: Séminaire de Probabilités IX. Proceedings 1973/74. Edité par P. A. Meyer. IV, 589 pages. 1975.

Vol. 466: Non-Commutative Harmonic Analysis. Proceedings 1974. Edited by J. Carmona, J. Dixmier and M. Vergne. VI, 231 pages. 1975.

Vol. 467: M. R. Essén, The Cos $\pi\lambda$ Theorem. With a paper by Christer Borell. VII, 112 pages. 1975.

Vol. 468: Dynamical Systems – Warwick 1974. Proceedings 1973/74. Edited by A. Manning. X, 405 pages. 1975.

Vol. 469: E. Binz, Continuous Convergence on C(X). IX, 140 pages. 1975.

Vol. 470: R. Bowen, Equilibrium States and the Ergodic Theory of Anosov Diffeomorphisms. III, 108 pages. 1975.

Vol. 471: R. S. Hamilton, Harmonic Maps of Manifolds with Boundary. III, 168 pages. 1975.

Vol. 472: Probability-Winter School. Proceedings 1975. Edited by Z. Ciesielski, K. Urbanik, and W. A. Woyczyński. VI, 283 pages. 1975.

Vol. 473: D. Burghelea, R. Lashof, and M. Rothenberg, Groups of Automorphisms of Manifolds. (with an appendix by E. Pedersen) VII, 156 pages. 1975.

Vol. 474: Séminaire Pierre Lelong (Analyse) Année 1973/74. Edité par P. Lelong. VI, 182 pages. 1975.

Vol. 475: Répartition Modulo 1. Actes du Colloque de Marseille-Luminy, 4 au 7 Juin 1974. Edité par G. Rauzy. V, 258 pages. 1975.

Vol. 476: Modular Functions of One Variable IV. Proceedings 1972. Edited by B. J. Birch and W. Kuyk. V, 151 pages. 1975.

Vol. 477: Optimization and Optimal Control. Proceedings 1974. Edited by R. Bulirsch, W. Oettli, and J. Stoer. VII, 294 pages. 1975.

Vol. 478: G. Schober, Univalent Functions – Selected Topics. V, 200 pages. 1975.

Vol. 479: S. D. Fisher and J. W. Jerome, Minimum Norm Extremals in Function Spaces. With Applications to Classical and Modern Analysis. VIII, 209 pages. 1975.

Vol. 480: X. M. Fernique, J. P. Conze et J. Gani, Ecole d'Eté de Probabilités de Saint-Flour IV–1974. Edité par P.-L. Hennequin. XI, 293 pages. 1975.

Vol. 481: M. de Guzmán, Differentiation of Integrals in R^n. XII, 226 pages. 1975.

Vol. 482: Fonctions de Plusieurs Variables Complexes II. Séminaire François Norguet 1974–1975. IX, 367 pages. 1975.

Vol. 483: R. D. M. Accola, Riemann Surfaces, Theta Functions, and Abelian Automorphisms Groups. III, 105 pages. 1975.

Vol. 484: Differential Topology and Geometry. Proceedings 1974. Edited by G. P. Joubert, R. P. Moussu, and R. H. Roussarie. IX, 287 pages. 1975.

Vol. 485: J. Diestel, Geometry of Banach Spaces – Selected Topics. XI, 282 pages. 1975.

Vol. 486: S. Stratila and D. Voiculescu, Representations of AF-Algebras and of the Group U (∞). IX, 169 pages. 1975.

Vol. 487: H. M. Reimann und T. Rychener, Funktionen beschränkter mittlerer Oszillation. VI, 141 Seiten. 1975.

Vol. 488: Representations of Algebras, Ottawa 1974. Proceedings 1974. Edited by V. Dlab and P. Gabriel. XII, 378 pages. 1975.

Vol. 489: J. Bair and R. Fourneau, Etude Géométrique des Espaces Vectoriels. Une Introduction. VII, 185 pages. 1975.

Vol. 490: The Geometry of Metric and Linear Spaces. Proceedings 1974. Edited by L. M. Kelly. X, 244 pages. 1975.

Vol. 491: K. A. Broughan, Invariants for Real-Generated Uniform Topological and Algebraic Categories. X, 197 pages. 1975.

Vol. 492: Infinitary Logic: In Memoriam Carol Karp. Edited by D. W. Kueker. VI, 206 pages. 1975.

Vol. 493: F. W. Kamber and P. Tondeur, Foliated Bundles and Characteristic Classes. XIII, 208 pages. 1975.

Vol. 494: A Cornea and G. Licea. Order and Potential Resolvent Families of Kernels. IV, 154 pages. 1975.

Vol. 495: A. Kerber, Representations of Permutation Groups II. V, 175 pages. 1975.

Vol. 496: L. H. Hodgkin and V. P. Snaith, Topics in K-Theory. Two Independent Contributions. III, 294 pages. 1975.

Vol. 497: Analyse Harmonique sur les Groupes de Lie. Proceedings 1973–75. Edité par P. Eymard et al. VI, 710 pages. 1975.

Vol. 498: Model Theory and Algebra. A Memorial Tribute to Abraham Robinson. Edited by D. H. Saracino and V. B. Weispfenning. X, 463 pages. 1975.

Vol. 499: Logic Conference, Kiel 1974. Proceedings. Edited by G. H. Müller, A. Oberschelp, and K. Potthoff. V, 651 pages 1975.

Vol. 500: Proof Theory Symposion, Kiel 1974. Proceedings. Edited by J. Diller and G. H. Müller. VIII, 383 pages. 1975.

Vol. 501: Spline Functions, Karlsruhe 1975. Proceedings. Edited by K. Böhmer, G. Meinardus, and W. Schempp. VI, 421 pages. 1976.

Vol. 502: János Galambos, Representations of Real Numbers by Infinite Series. VI, 146 pages. 1976.

Vol. 503: Applications of Methods of Functional Analysis to Problems in Mechanics. Proceedings 1975. Edited by P. Germain and B. Nayroles. XIX, 531 pages. 1976.

Vol. 504: S. Lang and H. F. Trotter, Frobenius Distributions in GL_2-Extensions. III, 274 pages. 1976.

Vol. 505: Advances in Complex Function Theory. Proceedings 1973/74. Edited by W. E. Kirwan and L. Zalcman. VIII, 203 pages. 1976.

Vol. 506: Numerical Analysis, Dundee 1975. Proceedings. Edited by G. A. Watson. X, 201 pages. 1976.

Vol. 507: M. C. Reed, Abstract Non-Linear Wave Equations. VI, 128 pages. 1976.

Vol. 508: E. Seneta, Regularly Varying Functions. V, 112 pages. 1976.

Vol. 509: D. E. Blair, Contact Manifolds in Riemannian Geometry. VI, 146 pages. 1976.

Vol. 510: V. Poènaru, Singularités C^∞ en Présence de Symétrie. V, 174 pages. 1976.

Vol. 511: Séminaire de Probabilités X. Proceedings 1974/75. Edité par P. A. Meyer. VI, 593 pages. 1976.

Vol. 512: Spaces of Analytic Functions, Kristiansand, Norway 1975. Proceedings. Edited by O. B. Bekken, B. K. Øksendal, and A. Stray. VIII, 204 pages. 1976.

Vol. 513: R. B. Warfield, Jr. Nilpotent Groups. VIII, 115 pages. 1976.

Vol. 514: Séminaire Bourbaki vol. 1974/75. Exposés 453 – 470. IV, 276 pages. 1976.

Vol. 515: Bäcklund Transformations. Nashville, Tennessee 1974. Proceedings. Edited by R. M. Miura. VIII, 295 pages. 1976.

Vol. 516: M. L. Silverstein, Boundary Theory for Symmetric Markov Processes. XVI, 314 pages. 1976.

Vol. 517: S. Glasner, Proximal Flows. VIII, 153 pages. 1976.

Vol. 518: Séminaire de Théorie du Potentiel, Proceedings Paris 1972–1974. Edité par F. Hirsch et G. Mokobodzki. VI, 275 pages. 1976.

Vol. 519: J. Schmets, Espaces de Fonctions Continues. XII, 150 pages. 1976.

Vol. 520: R. H. Farrell, Techniques of Multivariate Calculation. X, 337 pages. 1976.

Vol. 521: G. Cherlin, Model Theoretic Algebra – Selected Topics. IV, 234 pages. 1976.

Vol. 522: C. O. Bloom and N. D. Kazarinoff, Short Wave Radiation Problems in Inhomogeneous Media: Asymptotic Solutions. V. 104 pages. 1976.

Vol. 523: S. A. Albeverio and R. J. Høegh-Krohn, Mathematical Theory of Feynman Path Integrals. IV, 139 pages. 1976.

Vol. 524: Séminaire Pierre Lelong (Analyse) Année 1974/75. Edité par P. Lelong. V, 222 pages. 1976.

Vol. 525: Structural Stability, the Theory of Catastrophes, and Applications in the Sciences. Proceedings 1975. Edited by P. Hilton. VI, 408 pages. 1976.

Vol. 526: Probability in Banach Spaces. Proceedings 1975. Edited by A. Beck. VI, 290 pages. 1976.

Vol. 527: M. Denker, Ch. Grillenberger, and K. Sigmund, Ergodic Theory on Compact Spaces. IV, 360 pages. 1976.

Vol. 528: J. E. Humphreys, Ordinary and Modular Representations of Chevalley Groups. III, 127 pages. 1976.

Vol. 529: J. Grandell, Doubly Stochastic Poisson Processes. X, 234 pages. 1976.

Vol. 530: S. S. Gelbart, Weil's Representation and the Spectrum of the Metaplectic Group. VII, 140 pages. 1976.

Vol. 531: Y.-C. Wong, The Topology of Uniform Convergence on Order-Bounded Sets. VI, 163 pages. 1976.

Vol. 532: Théorie Ergodique. Proceedings 1973/1974. Edité par J.-P. Conze and M. S. Keane. VIII, 227 pages. 1976.

Vol. 533: F. R. Cohen, T. J. Lada, and J. P. May, The Homology of Iterated Loop Spaces. IX, 490 pages. 1976.

Vol. 534: C. Preston, Random Fields. V, 200 pages. 1976.

Vol. 535: Singularités d'Applications Differentiables. Plans-sur-Bex. 1975. Edité par O. Burlet et F. Ronga. V, 253 pages. 1976.

Vol. 536: W. M. Schmidt, Equations over Finite Fields. An Elementary Approach. IX, 267 pages. 1976.

Vol. 537: Set Theory and Hierarchy Theory. Bierutowice, Poland 1975. A Memorial Tribute to Andrzej Mostowski. Edited by W. Marek, M. Srebrny and A. Zarach. XIII, 345 pages. 1976.

Vol. 538: G. Fischer, Complex Analytic Geometry. VII, 201 pages. 1976.

Vol. 539: A. Badrikian, J. F. C. Kingman et J. Kuelbs, Ecole d'Eté de Probabilités de Saint Flour V-1975. Edité par P.-L. Hennequin. IX, 314 pages. 1976.

Vol. 540: Categorical Topology, Proceedings 1975. Edited by E. Binz and H. Herrlich. XV, 719 pages. 1976.

Vol. 541: Measure Theory, Oberwolfach 1975. Proceedings. Edited by A. Bellow and D. Kölzow. XIV, 430 pages. 1976.

Vol. 542: D. A. Edwards and H. M. Hastings, Čech and Steenrod Homotopy Theories with Applications to Geometric Topology. VII, 296 pages. 1976.

Vol. 543: Nonlinear Operators and the Calculus of Variations, Bruxelles 1975. Edited by J. P. Gossez, E. J. Lami Dozo, J. Mawhin, and L. Waelbroeck, VII, 237 pages. 1976.

Vol. 544: Robert P. Langlands, On the Functional Equations Satisfied by Eisenstein Series. VII, 337 pages. 1976.

Vol. 545: Noncommutative Ring Theory. Kent State 1975. Edited by J. H. Cozzens and F. L. Sandomierski. V, 212 pages. 1976.

Vol. 546: K. Mahler, Lectures on Transcendental Numbers. Edited and Completed by B. Diviš and W. J. Le Veque. XXI, 254 pages. 1976.

Vol. 547: A. Mukherjea and N. A. Tserpes, Measures on Topological Semigroups: Convolution Products and Random Walks. V, 197 pages. 1976.

Vol. 548: D. A. Hejhal, The Selberg Trace Formula for PSL (2, IR). Volume I. VI, 516 pages. 1976.

Vol. 549: Brauer Groups, Evanston 1975. Proceedings. Edited by D. Zelinsky. V, 187 pages. 1976.

Vol. 550: Proceedings of the Third Japan – USSR Symposium on Probability Theory. Edited by G. Maruyama and J. V. Prokhorov. VI, 722 pages. 1976.

Vol. 551: Algebraic K-Theory, Evanston 1976. Proceedings. Edited by M. R. Stein. XI, 409 pages. 1976.

Vol. 552: C. G. Gibson, K. Wirthmüller, A. A. du Plessis and E. J. N. Looijenga. Topological Stability of Smooth Mappings. V, 155 pages. 1976.

Vol. 553: M. Petrich, Categories of Algebraic Systems. Vector and Projective Spaces, Semigroups, Rings and Lattices. VIII, 217 pages. 1976.

Vol. 554: J. D. H. Smith, Mal'cev Varieties. VIII, 158 pages. 1976.

Vol. 555: M. Ishida, The Genus Fields of Algebraic Number Fields. VII, 116 pages. 1976.

Vol. 556: Approximation Theory. Bonn 1976. Proceedings. Edited by R. Schaback and K. Scherer. VII, 466 pages. 1976.

Vol. 557: W. Iberkleid and T. Petrie, Smooth S^1 Manifolds. III, 163 pages. 1976.

Vol. 558: B. Weisfeiler, On Construction and Identification of Graphs. XIV, 237 pages. 1976.

Vol. 559: J.-P. Caubet, Le Mouvement Brownien Relativiste. IX, 212 pages. 1976.

Vol. 560: Combinatorial Mathematics, IV, Proceedings 1975. Edited by L. R. A. Casse and W. D. Wallis. VII, 249 pages. 1976.

Vol. 561: Function Theoretic Methods for Partial Differential Equations. Darmstadt 1976. Proceedings. Edited by V. E. Meister, N. Weck and W. L. Wendland. XVIII, 520 pages. 1976.

Vol. 562: R. W. Goodman, Nilpotent Lie Groups: Structure and Applications to Analysis. X, 210 pages. 1976.

Vol. 563: Séminaire de Théorie du Potentiel. Paris, No. 2. Proceedings 1975–1976. Edited by F. Hirsch and G. Mokobodzki. VI, 292 pages. 1976.

Vol. 564: Ordinary and Partial Differential Equations, Dundee 1976. Proceedings. Edited by W. N. Everitt and B. D. Sleeman. XVIII, 551 pages. 1976.

Vol. 565: Turbulence and Navier Stokes Equations. Proceedings 1975. Edited by R. Temam. IX, 194 pages. 1976.

Vol. 566: Empirical Distributions and Processes. Oberwolfach 1976. Proceedings. Edited by P. Gaenssler and P. Révész. VII, 146 pages. 1976.

Vol. 567: Séminaire Bourbaki vol. 1975/76. Exposés 471–488. IV, 303 pages. 1977.

Vol. 568: R. E. Gaines and J. L. Mawhin, Coincidence Degree, and Nonlinear Differential Equations. V, 262 pages. 1977.

Vol. 569: Cohomologie Etale SGA 4¹/₂. Séminaire de Géométrie Algébrique du Bois-Marie. Edité par P. Deligne. V, 312 pages. 1977.

Vol. 570: Differential Geometrical Methods in Mathematical Physics, Bonn 1975. Proceedings. Edited by K. Bleuler and A. Reetz. VIII, 576 pages. 1977.

Vol. 571: Constructive Theory of Functions of Several Variables, Oberwolfach 1976. Proceedings. Edited by W. Schempp and K. Zeller. VI, 290 pages. 1977

Vol. 572: Sparse Matrix Techniques, Copenhagen 1976. Edited by V. A. Barker. V, 184 pages. 1977.

Vol. 573: Group Theory, Canberra 1975. Proceedings. Edited by R. A. Bryce, J. Cossey and M. F. Newman. VII, 146 pages. 1977.

Vol. 574: J. Moldestad, Computations in Higher Types. IV, 203 pages. 1977.

Vol. 575: K-Theory and Operator Algebras, Athens, Georgia 1975. Edited by B. B. Morrel and I. M. Singer. VI, 191 pages. 1977.

Vol. 576: V. S. Varadarajan, Harmonic Analysis on Real Reductive Groups. VI, 521 pages. 1977.

Vol. 577: J. P. May, E∞ Ring Spaces and E∞ Ring Spectra. IV, 268 pages. 1977.

Vol. 578: Séminaire Pierre Lelong (Analyse) Année 1975/76. Edité par P. Lelong. VI, 327 pages. 1977.

Vol. 579: Combinatoire et Représentation du Groupe Symétrique, Strasbourg 1976. Proceedings 1976. Edité par D. Foata. IV, 339 pages. 1977.